HAWAI'I'S FLORAL SPLENDOR

A FRIENDLY COLOR IDENTIFICATION GUIDE TO NATIVE AND INTRODUCED FLOWERS OF ALL THE HAWAIIAN ISLANDS.

ANGELA KAY KEPLER

MUTUAL PUBLISHING

(PREVIOUS PAGE) A rich, velvety hibiscus (*Hibiscus rosa-sinensis X schizopetale*) emerges to live one day only—on this occasion in the rain.

(LEFT) Brazilian bower plant (*Adhadota cydoniifolia*) occasionally graces humid lowland gardens.

First Printing September 1997
1 2 3 4 5 6 7 8 9

ISBN 1-56647-170-2

Mutual Publishing
1127 11th Avenue, Mezz. B
Honolulu, Hawai'i 96816
Telephone (808) 732-1709
Fax (808) 734-4094
Email: mutual@lava.net

Cover design by Mei Chi Chin
Book design by Angela Kay Kepler and Mark Abramson
All photos by Angela Kay Kepler unless otherwise noted

Printed in Taiwan

CONTENTS

INTRODUCTION

THE HAWAIIAN ISLANDS, stretching 1200 miles northwest from the Island of Hawai'i to Kure Atoll, half-way to Japan, comprise the most isolated major archipelago in the world. Distances to continents and other Pacific archipelagos are measured in several thousands of miles. Consequently, Hawai'i harbors an astonishing array of unique plants, and wildlife, both terraestrial and marine. Such evolutionary and ecological diversity is equally as remarkable as that found in the famed Galapagos Islands. Hawai'i is so distinct from every other island group in the world that biogeographers place it in its own ecological region.

Hawai'i harbors about 1000 species of native plants. Fully 89% of these are endemic, that is, are found nowhere else in the world. Unfortunately, more than 200 are officially endangered and many others are on a waiting list. When a plant is "federally listed," this generally means that only a handful are known... or that small numbers may be a factor precipitating its endangerment. For many extinct and endangered plants, only tattered shreds of their former glory remain: line drawings, brief descriptions, dried museum specimens, or perhaps an old photo. The Nature Conservancy, National Park Service, U. S. Fish and Wildlife Service, State of Hawai'i, Audubon Society, and many other private organizations are totally dedicated to the preservation of Hawai'i's native habitats. At the present rate of destruction, at least 80 remnant plant communities—half of the total number of original ecosystems known—will disappear within the next 20 years. Conservation activities are not only geared to endangered species and ecosystems, they include unrelenting efforts towards the sheer survival of Hawai'i's watersheds and coral reef ecosystems.

All six major Hawaiian islands are home to a richer variety of plants than one might expect. This is because Hawai'i's native flora is augmented by a plethora of introduced plants, mostly from tropical Asia and America. Since 1800, more than 5000 non-Hawaiian plant species have been introduced into Hawai'i. Most roadside, garden and landscaping beauty is not "truly Hawaiian" at all. To the uninitiated, "the islands" seem like "paradise." However, unbeknown to the casual observer, severe ecological imbalances have occurred during the past 200 years by deforestation, erosion, feral mammals, alien plant escapees, and exotic insects. Many "forest reserves" are either timber plantations, despoiled remnants of natural forests, or fenced and intensively managed, viable ecosystems.

On the other hand, the landscaping at resorts and in private botanical gardens is stunning and immaculate. Here the dominant flowers are of common occurrence throughout the world's tropics and subtropics.

Commercially, Maui and the Big Island (Island of Hawai'i) are major, international exporters of specialty flowers: heliconias, gingers, tropical foliage, proteas, carnations, anthuriums, orchids, and banksias. Maui is also the world center for protea research. These non-polluting industries help diversify the islands' economic bases, adding to their floral charm. The variety of top quality flowers has also propelled the well established art of flower arranging to new, distinctively neo-Hawaiian heights.

This book's 453 slides include 388 species and varieties of flowers, both native (70) and introduced (317) by early or recent immigrants. The 60 endemics (only found in Hawai'i)—whose entire geographic ranges often encompass tiny slivers of island habitat—are included to further assist the reader's appreciation of Hawai'i's renowned flora. Native plants are dotted in private and public gardens, National Parks and Preserves managed by The Nature Conservancy and State of Hawai'i. Don't forget annual festivals such as Lei Day, Kamehameha Day, Merrie Monarch Festival, etc.

HOW TO USE THIS BOOK

To facilitate identification, the book is divided into color-coded sections relating to flower color. Look for the color band at the top of each page. If a flower exhibits several color forms, it is cross-referenced in each pertinent chapter. Use common sense about colors: check for a magenta flower, for example, in both "Red" and "Purple".

Because of the "user-friendly" nature of this guide, I have attempted to organize chapters according to the dominant color of the flower or flowerhead. This is very important in species where there is more than one color, or where the main color is not due to true flowers but to colored, modified leaves (bracts). Examples are bougainvilleas, gingers, heliconias, calatheas, and other tropicals. The eye-catching purple, red, and orange bougainvilleas bear inconspicuous white flowers. Similarly, the six-inch flowerhead of golden beehive ginger bears small purple flowers which peek out shyly between the layered gold bracts (see "Yellow" chapter).

The colors of small greenish/whitish flowers are sometimes difficult to differentiate. Remember also that we may recall plant colors by their fruits rather than flowers. For example, in summer, before Christmas berry bushes develop the familiar scarlet berries, they bear masses of tiny green flowers. Since this is a book about flowers, Christmas berry occurs in the "Green" chapter.

NAMES

In Hawai'i, most native plants have Hawaiian names (italicized), as do *kama'aina* species introduced over a century ago. Some have no English names and are thus known solely by their botanical binomials.

PLANT ORIGINS

Each species or variety is coded according to its original home. In addition to the information's intrinsic interest, coding identifies plants of particular relevance to Hawaiian culture, biology, and conservation:

Flowers—like beautiful music—are best enjoyed live. Smell their sweet fragrances, drink in the visual artistry of their colors and patterns. Admire their biological adaptations for survival.

SYMBOLS FOR PLANT ORIGINS

- ▼ ABORIGINAL INTRODUCTION, I.E. BROUGHT BY EARLY POLYNESIANS
- ■ ENDEMIC TO HAWAI'I
- ◆ INDIGENOUS, I.E. NATIVE TO HAWAI'I BUT ALSO FOUND ELSEWHERE
- ✚ INTRODUCED

Finally, I wish to ask residents and visitors a special favor... please broadcast caring thoughts for the islands' natural ecosystems, all of which are priceless repositories of Hawai'i's natural heritage.

ACKNOWLEDGMENTS

GRATEFUL THANKS to all my hiking companions and scientific colleagues who shared their knowledge of, and enthusiasm for all Hawai'i's islands, from the Big Island to Laysan. Friends from the National Parks, U. S. and State Govenments, Sierra Club and Mauna Ala Hiking Clubs, The Nature Conservancy, University of Hawai'i, and commercial flower farms, will always be special. Particular gratitude is extended to Derral Herbst, Bob Hobdy, Mary Evanson, Cameron Kepler, and helicopter pilot, Tom Hauptman. Great thanks to the photographers who kindly assisted with color slides: Randy Bartlett, John Carothers, Haleakala National Park, Betsy Gagné, Cameron Kepler, Maui Inter-Continental Resort, Art Medeiros, Jacob Mau, Ron Nagata, and Eric Nishibayashi. I am also grateful to friends, too numerous to list, who showered me with aloha-filled hospitality during my resident and nonresident years, especially Roxie Berlin, the late Colin and Margaret Cameron, Anne Fielding and Bob Hobdy, Jonathan and Betty Mann, Jacob Mau and Mary Anne Doane-Mau, and my daughter, Leilani Kepler.

The floss silk tree's *(Chorisia speciosa)* flaring, rose-carmine flowers bloom profusely in the fall. In its native Argentina, this tree is dubbed *palo borracho* (drunken tree): its trunk, a water storage vessel, expands like a drunk man's belly.

Certainly one of the rarest trees in the world, Hawai'i's endemic and unusual *hau kuahiwi* (*Hibiscadelphus giffardianus*) is related to the common hibiscus. Extinct in the wild, it is only known from

ENVIRONMENTAL ALERT

Art Medeiros/Haleakala National Park

(ABOVE) Piles of burning purple plague (*Miconia calvescens*), Hana.

HAWAI'I, SITUATED AT 21° N latitude in the oceanic subtropics, is an ideal adopted home for introduced tropical plants. Most thrive in gardens and landscaping, presenting few problems. However, some escape into natural or semi-natural ecosystems and literally "run wild." The 1970s and 1980s experienced a proliferation of now-familiar pests such as strawberry guava (*Psidium cattleianum*); inkberry (*Ardisia elliptica*); yellow, white, and kahili gingers (*Hedychium spp.*); Java plum (*Syzygium cuminii*); and African tulip tree (*Spathodea campanulata*).

The 1990s ushered in more menacing species, two so virulent they are capable of *completely and rapidly wiping out Maui and the Big Island's lowland and upland forests.* The most lethal offender is *Miconia calvescens,* called purple plague or green cancer in Tahiti, where it has aggressively invaded forests from sea level to 4260 feet in less than 25 years. The other is cane tibouchina (*Tibouchina herbacea*).

Over 60% of the main island of Tahiti is now dominated by purple plague's dense, dark groves, which grow to 60 feet tall. Today, one-quarter of Tahiti's native plant species are nearly extinct.

Despite enormous expenses of time and money, Hawai'i's rainforests are poised for a similar scenario. Purple plague competes with feral pigs as the most serious threat to conservation in Hawai'i. Its potential impact is far greater than all other noxious plants combined, in particular, will render all past, present, and future conservation efforts futile, since it crowds out *all estab-*

WHAT YOU CAN DO ON MAUI, WHERE THE SITUATION'S WORST: PLEAS FROM NOXIOUS PLANT EXPERTS

1. **Whenever possible, destroy "purple plague" plants.** Look for them while driving around. Report sightings to the State Department of Agriculture, Pest and Weed Control Specialists (phone 871-5656) or Haleakala National Park (572-1983). Look particularly along the Hana Highway from Huelo to Kipahulu. Be specific: use stream names (check the concrete bridges) and landmarks (airports, addresses, waterfalls).

2. **No matter how pretty it may be, NEVER buy a purple-flowering bush with prominently 3- or 5-veined leaves.** To do so is a personal contribution to the guaranteed destruction of Hawai'i's rainforests, native biota, watersheds, and adjacent coral reefs. Alert nursery personnel and friends. Already 15 species of melastomes grow unchecked in Hawai'i, and almost all are listed as noxious weeds by the State of Hawai'i. Several common species are treated in this book and pictured opposite.

3. **Hikers: scrub boots and equipment after hiking anywhere on Maui or on other islands.**

Art Medeiros/Haleakala National Park

lished forms of life. Not even common birds, insects, and plants can survive its onslaught.

Purple plague's horticultural name is velvet leaf. In its native tropical America, where it colonizes light gaps, its populations are balanced by natural controls (insects, diseases) absent in Hawai'i where, under favorable conditions, a three-foot square experimental plot produced 18,000 seedlings in 6 months! Admittedly attractive, it is esteemed for its large (to two feet long), velvety leaves, shiny green above and bright purple below. Note the three **bold leaf veins**, a characteristic of its family, Melastomataceae.*

(TOP) Purple plague and Koster's curse (*Clidemia hirta*), see "White" chapter. The latter is already pestiferous on Oahu (250,000 acres of public and private lands) and in forests throughout Hawai'i. (RIGHT) Most of the greenery visible is purple plague, crowding out plants and animals unique to Tahiti. Maui and the Big Island are poised for identical situations.

(BELOW RIGHT) Cane tibouchina (*Tibouchina herbacea*), a new scourge on West Maui from Waihee to Honokowai (see "Purple" chapter). It is extremely important that this rampant, weedy intruder not be further transported by hikers to East Maui or to other islands.

(BOTTOM LEFT) Glory-bush (*Tibouchina urvilleana*) is a pest in Kokee, Kauai, and from Volcano to Glenwood, Big Island (see "Purple" chapter). Where it is a popular garden plant in moist areas, forests suffer irreparably.

(BOTTOM RIGHT) Medinilla (*Medinilla magnifica*), recently popular in nurseries throughout the islands, is already officially listed as a noxious weed in Hawai'i.

Randy Bartlett

*Some melastomes have five or seven bold veins, counting two close to the margin. Three, with many ladder-like cross-veins, is more typical.

BOTANICAL GARDENS, PARKS, AND PRESERVES IN HAWAI'I

Hikers: In order to prevent the spread of noxious weeds, please—ALWAYS—scrub boots and equipment after hiking anywhere.

On account of extensive private and state lands throughout the islands, and problems associated with illegal marijuana growing, permission is necessary to hike almost anywhere outside the following established public areas. Exceptions are National Parks (prior permission is necessary for overnight cabins), and State and County Parks. Remember that illegal hiking carries with it the extra hazards of confrontation with illegal marijuana growers and pig or goat hunters.

The following list is incomplete. Check the yellow pages or write to State of Hawai'i for maps of State Parks, P.O. Box 621, Honolulu, HI 96809, The Nature Conservancy, 1116 Smith St., Honolulu, HI 96817, or individual National Parks. Remember that throughout Hawai'i are flowers: family gardens, commercial landscaping, country roads, etc. Native forest is limited, generally requiring visiting special reserves except parts of Hawai'i and Kauai.

Some nurseries offer tours, especially if you show interest in buying or shipping plants. Nurseries may be general or specialize in ferns, bromeliads, orchids, anthuriums, proteas, tropicals (gingers, heliconias), etc.

ISLAND OF HAWAI'I ("BIG ISLAND")

The Big Island has more accessible native forest than any other Hawaiian island, including an outstanding national park, many state parks and recreation areas, and many commercial gardens with tropical flowers for sale. If you are interested in heliconias, gingers, or orchids, the Big Island and Maui are best. The former is best for anthuriums.

AKAKA FALLS STATE PARK: public, no fee, 65 acres, pleasant, self-guided walk, lush tropical vegetation, beautiful falls (420 feet high) and Kahuna Falls (400 feet).

AKATSUKA ORCHID GARDENS: private, no fee, commercial, 100s of species. Located on Highway 11, 3 miles east of Volcano Village, near the National Park.

HAWAI'I TROPICAL BOTANICAL GARDEN: private, fee, 17 acres, lush lowland, "jungly" garden with introduced tropical gingers, heliconias, orchids, palms, etc. Located on the 4-mile Scenic Route off Highway 19 north of Hilo, just north of Hokeo Point. Walking trails, many plants labelled.

HAWAI'I VOLCANOES NATIONAL PARK: public, free, 207,643 acres, habitats: cinder desert to lush rainforest. Camping (permits for backcountry cabins), elevation sea level to 4000 feet, Crater Rim Drive and excellent hiking trails feature lava flows and tubes, volcanic craters, forests, steam vents, sulfur banks. Not to be missed.

LILIUOKALANI GARDENS: public, no fee, leisurely, Japanese, urban park, Banyan Drive, close to resorts on southwest Hilo Bay.

KALOPA NATIVE FOREST STATE PARK: public, natural park in the very large wilderness, Hamakua Forest Preserve, free, 100 acres, elevation 2000 feet, 100 inches rain p.a. Campsites, picnics, huts, small arboretum, and Native Forest Nature Trail with brochure. At end of Kalopa Road, 3 miles inland from Mamalahoa Highway (Route 19), 5 miles southest of Honokaa, north of Hilo.

KAUAI

KEAHUA FORESTRY ARBORETUM: public, free, in Waileale Forest Reserve. Located north of Kokee, drive inland on Highway 580 from Wailua to end of road. Walking trails through eucalyptus/native forest. Picnic, swimming pool.

KILAUEA POINT NATIONAL WILDLIFE REFUGE: public, small fee, special haven for seabirds (Red-footed Booby, Wedge-tailed Shearwater, Great Frigatebird, etc.), in a native coastal plant environment. Most northerly promontory, scenic.

KOKEE STATE PARK: public, free, 4640 acres, elevation 3600 feet, camping. Beautiful montane park, with outstanding views of north coast from trails. Native and introduced trees, Hawaiian endemics such as native begonia (*pua-maka-nui*), *'ohi'a, koa.* Located near end of Waimea Canyon Road (Highway 55), Northwest Kauai.

NA PALI COAST STATE PARK: public, no fee, 6500 acres, hikers only. Spectacular coastal vistas, backcountry tenting, sea cliffs, lush valleys, introduced and native coastal plants, 11 miles to Kalalau Valley (trail end).

NATIONAL TROPICAL BOTANICAL GARDEN: (= Pacific Tropical Botanical Garden). Private, fee, 186 acres, labelled plants. Kauai's best. Advance reservations necessary NTBG, P.O. Box 340, Lawai, Kauai, HI 96765. Rare and endangered tropical plants, incl. Hawai'i. In Lawai, Highway 50, south shore east of Kalaheo.

OLA PUA BOTANICAL GARDENS: private, fee, long established garden with abundant flowering plants. Look for sign to a gravel road off Highway 50, just west of Kalaheo, south shore (next to a big macadamia nut orchard).

WAIMEA CANYON STATE PARK: public, free, 1866 acres. Spectacular views of a "Grand Canyon"-like eroded, deep gorge; lovely native *koa* trees, endemic palm-like *iliau,* etc. Strenuous hikes into canyon, which harbors rare plants.

MAUI

ALI'I GARDENS: private, fee, 7 acres, lush tropical vegetation, gingers, heliconias, foliage. Ocean view, Hana Highway at Nahiku west of Hana airport.

HALEAKALA NATIONAL PARK: public, fee, elevation SL to 10,020 feet, 43 square miles, World Heritage Site. Native subalpine vegetation, lunar scenery, trails, cabins inside wilderness area, top of Route 378 (37 miles from airport). Kipahulu: lush lowland vegetation, spectacular coastal and mountain scenery, trails, waterfalls—70 miles along north coast's Hana Highway from airport.

HANA GARDENLAND: private, fee, 5 acres, nursery plus tropical garden, includes cultural specialties, e.g. *'awa;* ships plants worldwide; 200 feet elevation.; just west of Ulaino Road and Hana airport, 50 miles east of Kahului.

HELANI GARDENS: private, fee, 70 acres, sea level to 600 feet elevation, lush foliage, esp. gingers, heliconias; trails or drivethrough. On east side, take Highway 360 (Hana Highway) from Kahului to Hana, 1 miles north of Hana on right.

IAO VALLEY STATE MONUMENT: public, free, 6 acres; lush lowland vegetation, streams, spectacular ridges, valleys and mountain peaks; trails; end of Iao Valley Road (Route 32), west of Kahhului and Wailuku.

KAHUNA GARDENS: private, fee, 120 acres, owned by National Tropical Botanical Garden, often no-one attending. Specialty: Polynesian ethnobotany, open 10-2 Tuesday-Saturday, call (808) 248-8912. Located 1.5 miles makai (towards the sea) of Hana Highway (Route 360) on Ulaino Road, just west of Hana airport.

KEANAE ARBORETUM: public, free, 6 acres, maintained by State of Hawai'i, cultural plants, including taro; tropical plants worldwide, some native, most labelled. Located just west of Keanae village, Hana Highway (Route 360).

KEPANIWAI HERITAGE GARDENS, IAO VALLEY: public, free, 8 acres; for picnics, emphasizes major ethnic groups in Hawai'i with matching architecture and plants (Chinese, Japanese, Portuguese, etc.); en route to 'Iao Valley State Monument.

KULA BOTANICAL GARDEN: private, fee, 5 acres, elevation 3,300 feet; big variety introduced and native plants (big *koa* trees), proteas; stunning views. Located on slopes of Haleakalaa, 1/2 mile from the junction of Kekaulike Road and Kula Highway (Route 37). Maui's best maintained botanical garden.

MAUI BOTANICAL GARDEN: public, free, County of Maui, sea level; native Hawaiian plants and Polynesian introductions (big *kamani, kukui,* and breadfruit); in Kahului Zoo, Kanaloa Road., near Kaahumanu Shopping Center.

MAUI TROPICAL PLANTATION: private, popular with tours, tropical fruits and flowers (heliconias, etc). In West Maui foothills, south of Wailuku (Route 30).

TROPICAL GARDENS OF MAUI: private, fee, 4 acres, lush tropicals. Iao Valley Road near Kepaniwai Park, west of Kahului. Open Mon.-Sat. 9-5. Ships orchids.

UNIVERSITY OF HAWAI'I EXPERIMENT STATION: public, free, 20 acres, specializes in proteas; Mauna Place (off Copp Road, Kula), 3,200 feet elevation; self-guided tour with map. Opposite is Proteas of Hawai'i, available for informal tours.

WAILEA POINT COASTAL TRAIL: public, free, 1/2 miles trail at south end of Wailea Beach, with rugged lava. Superb native plants, some very rare.

WAIKAMOI PRESERVE: private, call The Nature Conservancy, 5,230 acres, elevation 4400—8600 feet. Abuts Haleakala National Park and Hanawi Natural Area Reserve (total >100,000 acres of rain forest), harbors 14 ecosystems, common and rare flora and fauna. No public access. Maui's most accessible, prime rain forest.

MOLOKAI

HALAWA VALLEY TRAIL: free public trail, drive south coast east to last valley, Halawa. Trail, to 2 falls, lush lowland forest; may close due to floods.

KAMAKOU PRESERVE: private, free, 2774 acres, native montane rainforest. Excellent hiking trails, magnificent views, hikers appropriately dressed only! Reserve through The Nature Conservancy, P. O. Box 40, Kualapuu, HI 96757.

MOOMOMI BEACH PRESERVE: private, free, native coastal plants. Reserve through The Nature Conservancy (above), on northwest shore. Excellent, scenic.

OAHU

Best botanical gardens, city landscaping, and educational programs in state.

FOSTER BOTANIC GARDEN: private, small fee, 20 acres, est. 1855, educational programs, trails, map, booklet, labelled plants of 1000s of species, Hawai'i and worldwide. On Vineyard Boulevard between Liliha Street and Nuuanu Avenue, Honolulu, near downtown. 9-4 daily. Excellent. Also 24 of Hawai'i's "exceptional trees".

HOOMALUHIA PARK: public, free, 400 acres. Botanic garden and nature preserve, educational programs, Hawaiian ethnobotany, crafts, hiking, camping. Open daily 9-3. Located at end of Luluku Road, off Likelike Highway, one mile inland of Kamehameha Highway, near Windward City Shooping Center, Kaneohe, windward Oahu.

KAENA POINT NATURAL AREA RESERVE: hiking trail, coastal plants, best shoreline hike on Oahu, est. 1983, arid, part of Hawai'i Natural Area Reserve System. Located at end of road, far west point of Oahu. Take H1 freeway from Honolulu west to west shore, where it turns into Farrington Highway. (930).

KOKO CRATER BOTANIC GARDEN: public, free, always open, arid, with succulents and dryland plants, many Hawaiian endemics. Take Kalanianaole Highway (Route 72) east from Waikiki, past Hanauma Bay to Koko Head Sandy Beach Park. Turn left into Queens Gate subdivision, Kealahou Street to Koko Head Stables.

LYON ARBORETUM: public, free, 124 acres, >4000 species, lush, part of Univ. of Hawai'i. Educational programs, trails, views, Hawaiian ethnobotany, native plants, many old trees. Open daily 9-3. Take Punahou Street or University Avenue, East Manoa Road, Oahu or Manoa to upper Manoa Valley past Paradise Park.

WAHIAWA BOTANIC GARDEN: public, free, 27 acres, est. 1950 but a forestry experiment station since the 1920s. Lush, 100s of species, trails. In central Oahu: take H2 freeway to Wahiawa, where it turns into Kamehameha Highway, take right on 3rd street (California Ave), garden on left down a few blocks.

WAIMEA ARBORETUM: private, fee, large, >3000 species, including many endangered Hawaiian plants. Excellent, next to Waimea Falls Park. Trails, labelled plants, conservation oriented. On northern Waimea Bay, on Kamehameha Highway (Highway 83): take either H2 freeway (central Oahu) or go around east shore.

■ 'Opelu (*Lobelia grayana*), a stunning native lobelia which now numbers in the hundreds of individuals in Haleakala National Park since goats have been eliminated.

BLUE FLOWERS

PURE, UNADULTERATED BLUE is an uncommon color in nature, especially in the tropics. It occurs primarily in wildflowers native to temperate North America and Europe (forget-me-nots, bluebells, gentians). Few true "wildflowers" occur in Hawai'i. Roadside flowers—rarely blue—are generally weedy imports from elsewhere. This chapter details a few uncommon ornamentals and native plants, mostly from Maui's upcountry. Both humans and bees register "blue" flowers as purple, lavender, or other hues within the blue-violet light spectrum, so consult the "Purple" chapter also. Curiously, blue flowers are rarely fragrant.

(RIGHT) A mixed blue/purple hat lei mingles blue hydrangea with the lustrous foliages of silver tree (*Leucadendron argenteum*) and a cultivated wormwood (*Artemisia sp*).

✚ AGAPANTHUS LILY
(*Agapanthus praecox*) Liliaceae

(CENTER & RIGHT) Also called African lily or lily-of-the-Nile, this showy perennial accents sunny, upcountry gardens. Native to South Africa, its underground bulb thrusts up tall stalks (to three feet) bearing clusters of bluish-purple, tubular blossoms. In leis, here nestled among fuchsias, they last well **(BELOW)**. Agapanthus flowers (blue or white), are occasionally seen at weddings—an appropriate use since their scientific name translates as "African love flowers".

Opposite: Ron Nagata/Haleakala National Park

✚ CALATHEA, ICE-BLUE
(*Calathea burle-marxii cv. 'Ice Blue'*)
Marantaceae

(RIGHT) In real life, this ethereal flowerhead resembles glacial ice ... its delicate translucency laced with darker blue veins must be seen to be fully appreciated. Introduced to Hawai'i in 1973, 'Blue Ice' and its close relative, 'Green Ice' (see "Green" chapter) remained obscure until tropical flowers became commercially popular. Note the mauve flowers emerging from behind the cupped bracts. This crystalline floral jewel, only recently described as a new species, grows best in shady, moist, humid areas.

✚ HYDRANGEA
(*Hydrangea macrophylla*)
Hydrangaceae

(LEFT) A familiar mainland ornamental, hydrangeas may be spotted in the higher upcountry gardens. The original Japanese species was introduced very early into English horticulture (1790). The flowers grow in dense, rounded clusters several inches in diameter. Their tiny, buttonlike centers often lack reproductive parts, rendering them sterile. Although hydrangeas are typically north temperate in origin, Hawai'i is unusual in that it has an endemic species, the *kanawao* (see "White" chapter).

■ *'OPELU*, AN ALPINE LOBELIA
(*Lobelia grayana*) Campanulaceae

(RIGHT) *'Opelu* (pron. "opay-lu") is a sparingly branched specialty of Haleakala's alpine shrublands, its entire world home. It frequents rocky or cindery outcrops, thrusting up tall, crowded flowerheads of curved, blue-violet flowers reminiscent of aloes (see "Red" chapter). *'Opelu* is unmistakable even when not blooming: its leaves are long, narrow, and so densely hairy on the underside they glisten brilliantly in the sunshine. This leaf shape and radiance must have inspired the ancient Hawaiians to call the plant 'opelu, after its marine "twin," a pelagic jackfish (*Decapterus pinnulatus*) with a narrowly elliptical shape and silvery color. Silvery-leaved plants always indicate an adaptation to high elevation or arid environments.

Ron Nagata/Haleakala National Park

◆ *PA'U-O-HI'IAKA*, A COASTAL VINE
(*Jacquemontia ovalifolia*) Convolvulaceae

Legend recounts that once upon a time Pele, the famous Hawaiian fire goddess, went fishing and left her baby sister Hi'iaka, asleep on the sand. Rushing back later than expected, Pele was relieved to find that a beach vine had blanketed the baby, protecting her from the midday heat. Ever since then, *Jacquemontia* has been called *pa'u-o-Hi'iaka*: Hi'iaka's skirt.

(RIGHT) This lowly vine (pron. "pah-ew-oh-hee-ee-ah-kah"), with pallid blue or white, bell-shaped flowers, favors arid, sandy or rubbly beaches. Look for it at Wailea Point (Maui), Kaena Point (Oahu), & Moomi Beach (Molokai).

(LEFT) *Pa'u-o-Hi'iaka* creeps among driftwood. Some biologists believe that its first seeds were carried by wind-tossed logs originating in tropical America. Driftwood originating from mainland America is very localized in Hawai'i: it funnels into a handful of coves, primarily on Maui and Kahoolawe.

✚ PLUMBAGO
{*Plumbago auriculata* (=*capensis*)} Plumbaginaceae

Native to South Africa, this climbing, perennial shrub tolerates both heat and cool and often forms hedges **(RIGHT)**. Its curious names—plumbago and leadwort—derive from an ancient European belief that the leaves cured lead poisoning.

(LEFT) Periodic heavy pruning triggers a fresh proliferation of its five-petalled flowers (1/2 inch across). The sepals (under the petals) are characteristically sticky. An indigenous plumbago (*P. zeylanica*), whose sap was used to blacken tattoos, grows in dry habitats.

✚'Green Ice' calathea (Calathea cylindrica), a Brazilian rainforest jewel, can be grown to at least 2000 feet in humid, shady nooks.

GREEN FLOWERS

FLOWERS IN NATURAL ENVIRONMENTS do not exist for us to enjoy ... they have evolved to reproduce themselves, to be pollinated. That's all. It's the horticulturalists that spend time and energy on our account! The most well-known pollinators are birds, bats, and insects, but wind is also important, especially for green flowers. Granted, they are often uninteresting, being tiny or catkin-like, with no colors for easy identification. Some are little more than tufted stamens bristling among the leaves. To the wind, bright colors are superfluous! Despite this, Hawai'i harbors a splendid array of truly noteworthy green flowers, some exhibiting clever pollination strategies, most by birds or insects rather than wind. They hide in lei shops, protea bouquets, gardens, heliconia farms, and deep within montane forests.

(TOP RIGHT) Elegant cymbidium orchids (*Cymbidium hybrids*) accentuate the boldness of yellow caribaea (*Heliconia caribaea* cv. 'Cream'), an erect heliconia.

■ *ALA-ALA-WAI-NUI*, FOREST "PEPPERS" (*Peperomia*) Piperaceae

Hawai'i harbors 25 native species of *ala-ala-wai-nui* (pron. "allah-allah-why-noo-ee"), fleshy herbs which frequent moist forests. They share the same family as commercial pepper. Their upright green flower spikes are composed of hundreds of tiny, perfect flowers or rounded fruits. **(RIGHT)** A tall species from Iao Valley and **(FAR RIGHT)** a shorter, red-stalked species from Waikamoi Preserve, Maui.

✚ ANTHURIUMS (*Anthurium andraeanum*) Araceae

Wild anthuriums, abundant in tropical American rain forests, are green. Although anthuriums are now associated with Hawai'i, they are not native to the islands; the original anthurium was brought from Colombia in 1889 by an English missionary, Samuel Damon. **(RIGHT)** When colored anthuriums age on the plant, they usually turn green. Hybridization is accomplished by hand pollination, monitoring of seeds from the knobby central column (spadix), and careful nurturing for two or three years.

Green-colored hybrid anthuriums are often called *midori*, Japanese for green. Many cultivars bear Japanese or Hawaiian names. When grown indoors, plants benefit by frequent misting and fertilizing. To ensure good drainage, place pots in trays of pebbles and water.

✚ BANKSIAS
(*Banksia*) Proteaceae

Stiffly elegant and long-lasting, banksias are instantly recognizable: they are *different*, resembling spiky corn cobs. Each flowerhead is composed of hundreds of crowded, spirally arranged, curiously shaped flowers. All banksias (71 species) originated in arid habitats of Australia. To early explorers and settlers, they were so conspicuous almost all were discovered before 1800. Approximately 20 species are grown commercially on the slopes of Haleakala between 3000 and 4000 feet. In addition to commercial benefits, Maui's flower farms provide outdoor reserves for species dwindling in their natural habitats.

(ABOVE) A popular cut flower is the rickrack banksia (*Banksia speciosa*), whose foliage resembles sewing rickrack. It lasts days without water, an obvious shipping advantage.

(LEFT) 'Summer Lime' (*Banksia baxteri*) is a furry, desert-adapted banksia. It too, sports rickrack leaves, but they are coarser than B. speciosa and appear to have been chopped off at their tips.

▼ BREADFRUIT OR *'ULU*
(*Artocarpus altilis*) Moraceae

Polynesians recount many tales of the origin of mankind: one "lineage" traces back to breadfruit, in part because of the fruit's resemblance to a human head. Rich in Pacific culture, *'ulu* was so essential to early Hawaiian colonists (food, medicines, wood, etc.) they brought it over 2000 miles in wave-washed canoes—not an easy task, since the principal variety only grows from suckers, is not very salt-tolerant, and there were no plastic bags in those times. Some old-timers (kama'aina) still make tasty "potato" salads with *'ulu.* The numerous female flowers are all fused together in a miniature version of the globular fruit, while the tiny male flowers are yellow, fused into a cylindrical spike. Look for its huge, glossy, lobed leaves, especially along windward coasts.

✚ CHRISTMAS BERRY
(*Schinus terebinthifolius*) Anacardiaceae

Also known as Brazilian pepper tree, this widespread shrub is slightly poisonous; some people suffer allergenic skin and respiratory problems similar to poison ivy. Nevertheless, its bountiful clusters of red berries are popular in Hawai'i for Christmas decorations. Look for it in pastures (where it is a pest), along secondary roads, and in gardens. Its local name, wilelaiki, honors Willie Rice, a politician whose legacies included the wearing of Christmas berry hatbands. In summer, it bear masses of tiny greenish flowers.

✚ "FEATHER" OR "PINEAPPLE" CALATHEA
(*Calathea sophia*) Marantaceae

(RIGHT) Softly scaly, this five-inch beauty is composed of white-tipped green bracts, overlapping in bud to resemble a feather. A newcomer to Maui in the 1990s, it hails from warm, humid, South American rainforests.

(ABOVE) On maturing, "feather" calathea fattens and spreads its crown, transforming into a fleecy "pineapple." Look closely to find the purple flowers emerging from the bracts. Other species of calatheas and marantas bear strikingly patterned leaves, and have recently become popular houseplants.

✚ GREEN HELICONIA
(*Heliconia indica var. rubricarpa*)
Heliconiaceae

Although most of the gaudy heliconias originated in tropical America, the greens are more typical of Asian and Western Pacific rainforests. None reached Hawai'i on their own. The 6 species of green-flowering heliconias, ranging from Samoa to Indonesia, are utilized more culturally than the entire 250 species in tropical America! Their ten-inch flowerheads, less visually appealing than familiar heliconias, are fascinating in that they are pollinated by nectar-sipping bats instead of hummingbirds (which are only found in the Americas). Since bats are color-blind, there has been no evolutionary pressure to develop color in either the flowers or flowerheads. The smooth shininess and apple-green hue of green heliconias are best appreciated in combination in mixed bouquets.

✚ JADE VINE
(*Strongylodon macrobotrys*) Leguminosae(Papilionaceae)

Deep within Luzon rain forests (Philippines), a rampantly twining liana evolved into one of the most spectacular and strange-colored flowers in the world. Soft and thickly textured, the blue-green, "beaked" blossoms (three inches long) are borne in long clusters. Long-lasting, they can be flattened and cleverly woven (*mauna-loa style,* see "Pink" chapter) into cherished leis.

■ *MAMAKI*, A FIBER BUSH
(*Pipturus albidus*) Urticaceae

Mamaki is familiar to most lowland hikers, since it is common along streambeds and beside waterfalls, mostly between 1500 and 4000 feet. Its toothed leaves, pale beneath, are broadly egg-shaped, often with pink petioles and veins. The 3 major veins stand out amid a prominent network of smaller veins. The flowers, arranged in small clusters along the stems, are tiny and greenish (male) or pinkish (female). A non-stinging member of the nettle family, *mamaki* was used to make a poorer grade barkcloth (tapa) and in medicines. Even today, the bark is used in tonics, the leaves in herb teas, and the fruit for a mild laxative effect.

■ *'OHA-WAI,* TREE LOBELIAS
(*Clermontia*) Campanulaceae

John Carothers

Hawai'i's native tree lobelias hold international esteem. To biologists, *'oha-wai* (pron. "oh-hah-vy"), *haha,* and kin exhibit prime examples of evolution in isolation. Most of the 112 species have highly restricted geographic ranges (see other chapters). The Nature Conservancy Preserves are good places to see a few, remembering that nearly 25% of the species are extinct.

(LEFT) Sharp-eyed plant-watchers may spot *Clermontia kakeana* along the forest margins of the Hana Highway at Nahiku and above the Boy Scout camp at Waihee, West Maui. *'Oha-wai's* green flowers are unusual in that they are pollinated by Hawai'i's own endemic honeycreepers.

▼ *'OLENA* OR TURMERIC
(*Curcuma longa*) Zingiberaceae

Until recently, it was difficult to find *'olena* in Hawai'i, despite its ethnobotanical history. Thanks to the commercial trade in tropical flowers, it is now available for growing and as a cut flower. This beautiful ginger (similar to its relatives, the calatheas), was used in old Hawai'i to dye long yellow skirts (*pa'u*) and as a ceremonial purifier of people and things. Only the finest white barkcloth was used to ensure a bright, pure color. The bright yellow, powdery dye was simply dried, powdered rhizome, identical to the turmeric we buy today to flavor curries. Indeed, the scientific name, *Curcuma,* derives from the Arabic *kurkum,* which describes the exact yellow color of *'olena's* roots and flowers.

✚ POINSETTIA
(*Euphorbia pulcherrima*) Euphorbiaceae

Although we associate poinsettias with red color at Christmas, the plants are green for most of the year. Their bright "petals" are bracts (modified leaves) and the true flowers are bunched in greenish, button-like clusters. Poinsettias are called "photoperiod plants" because their chemistry responds to changing day length: shorter days and longer, cooler nights stimulate the development of red leaf pigments. In Hawai'i, they need pruning biannually to produce red leaves: once after the blooming season and again in August. Commercial growers spray plants with dwarfing chemicals to produce the small potplants available in stores at Christmas.

✚ PROTEAS
(*Protea*) Proteaceae

(BELOW) The apple-green protea (*P. coronata*), with its clear-green tissues, crowned with a mound of white fleece, is especially beautiful when young. Like green heliconias, apple-green proteas are best appreciated in mixed flower arrangements, where they have a remarkable ability to balance colors. In shape, the flowerhead resembles the princess protea (*P. grandiceps*). Both are neatly compact, and generously endowed with "fur" to insulate them from cold winds, a legacy of their original alpine habitats.

(BELOW) This photo illustrates two green proteas, the apple-green on the right and a greenish-yellow form of pink mink (*P. neriifolia,* see "Pink" chapter) on the left. Protea flowerheads are complex, but their make-up is similar to an artichoke. The fleshy, layered edible portions of the artichoke are also bracts, and the tight, spiky filaments that you scrape of the "heart" are its mass of flowers, so modified that only basic reproductive structures remain.

▼ TI, A "CABBAGE TREE"
(*Cordyline fruticosa*) Liliaceae

Although ti's true flowers are small and pink ("Pink" chapter), this notable Hawaiian plant is included here because in recent years island lei-makers have become skilled in the art of ti-leaf "flowers" **(CENTER)**. The broad, strap-like, shiny leaves **(TOP)**, picked from commercial or private gardens, or from the wild, are cleverly folded and spiralled into the Hawaiian equivalent of Japanese tomato "roses". Ti (correctly *ki*, but no-one calls it that), transported carefully by Polynesians over thousands of miles of trackless ocean, was an indispensible item in early Hawaiian culture. It provided food wrappers, fermented "beer", woven sandals, a sweetish starch, and a truce symbol. Its uses may have changed today, but it remains a true symbol of the "real" Hawaiians.

✚ "WHITE SNAKE"
(*Calathea playstachys*) Marantaceae

Another botanical curiosity and newcomer to Hawai'i in the 1990s, "white snake" originated in Central American rainforests. I have only seen it growing on Bob Getzen's farm in Ulaino, Maui, where he cannot keep up the demand for it. "White snake's" odd flowerhead, about six inches long, is plainly two-dimensional and thinner than your little finger. Its scaly, serpent-like bracts interweave in a remarkably even pattern. Note the little orange flowers, appearing in ones and twos. Its closest relative is red or yellow, locally called rattle shaker, shakers, or rattlesnake plant (*Calathea crotalifera*).

Commercially known as 'Sassy', this spectacular parrot's beak heliconia (Heliconia psittacorum cv. 'Kaleidoscope') makes a distinctively exotic ground cover, since it blooms all-year.

ORANGE FLOWERS

ORANGE IS BOLD. Its wavelengths, spanning the spectrum from fiery red-orange to apricot—always magnetize. Most of Hawai'i's showy orange flowers originated in the continental tropics, where they evolved bright colors in order to attract bird and insect pollinators (especially butterflies). Although none of their natural pollinators live in Hawai'i, in some species local birds and insects have taken over their role.

(RIGHT) A melange of orange and complementary colors comprise an unusual *haku* lei, including orchids, cup-and-saucer plant, yellow *'ohi'a lehua,* iceplant, leather fern, and the fern relative, moa (*Psilotum nudum*).

✚ AFRICAN TULIP TREE
(*Spathodea campanulata*) Bignoniaceae

(BELOW) Globes of frilly, tulip-shaped flowers, resembling lopsided cups of molten steel, protrude boldly from this tall tree's dark, compound-leaved foliage. Budding off from a central mass of curved brown crescents, circles of these asymmetrical, scarlet-orange blossoms appear year-round, especially from January to June. On Maui it grows up to about 3000 feet.

Both blossoms and leaves yield attractive dyes ranging from mustard to dark brown. The African tulip tree, originally from Africa, is unrelated to "real" tulips or tulip poplars: its kin are trumpet vines and jacarandas. Its elliptical, woody seedpods furnish dry material for flower arangements, while its papery, silvery winged seeds may be fashioned into fluffy leis **(RIGHT)**.

Maui Inter-Continental Resort

✚ ANGEL'S TRUMPET TREE

{Brugmansia (=*Datura*) *candida*} Solanaceae

(RIGHT) *Nanahonua,* the angel's trumpet, with its enormous, bell-shaped, pendulous flowers (remember His Master's Voice gramaphone speakers?), is unmistakable. A white variety is described in the "White" chapter. This small tree thrives in gardens at all elevations, and occasionally dots the windward lowland forests. Originally from the Andes, its sap contains toxic alkaloids including scopalomine (used in anti-motion sickness ear patches) and hallucinogens. *All parts of this plant are poisonous.*

✚ BIRD-OF-PARADISE

(*Strelitzia reginae*) Strelitziaceae

Cheery and boldly curvaceous, the bird-of-paradise flowerhead resembles a golden crested crane. Arising from a boat-shaped basal sheath, up to six dazzling flowers emerge over a period of two weeks in a sunburst of glossy orange and blue. Carefully coax extra flowers from the "boat" for a fuller look to your flower arrangements. Seeds rarely develop, since birds-of-paradise are cleverly designed to be pollinated by African sunbirds, not present in Hawai'i. Stylized "birds" can be found in art: paintings, stained glass, silk clothing, quilts.

✚ BLACK-EYED SUSAN VINE

(*Thunbergia alata*) Acanthaceae

This unmistakable, gay ground cover, originally from tropical Africa, is found in gardens and is locally common en route to Haleakala National Park (near Kula Lodge). It lacks the trumpet shape of most *Thunbergias,* and the overlapping bracts of typical members of the Acanthaceae. Nevertheless, it is a delightful drought-adapted climber (stems to 10 feet) that is easily raised from seed; it is surprising that more people do not grow it.

✚ BOUGAINVILLEA
(*Bougainvillea*) Nyctaginaceae

Symbolic of tropical brilliance, bougainvillea grows most prolifically in hot sunny lowlands. Its striking colors are due to modified *leaves* (bracts) rather than *flowers*. In its native Brazil, these bursts of color (originally purple) evolved to attract hummingbirds. An unusual lei can be fashioned by passing a needle and thread through small clusters of the colored leaves. For maximum color, bougainvillea requires heavy annual pruning, so be careful of its vicious spines! See other chapters also.

✚ BROMELIADS
(*Guzmania*) Bromeliaceae

The approximately 2000 species of bromeliads, native to tropical America, have been cultivated since the 1880s. Stiffly rosetted, these pineapple relatives have become increasingly available in Hawai'i in recent years. Look for them in hotel lobbies and private lowland gardens. **(MIDDLE RIGHT)** 'Scarlet star' (*Guzmania lingulata*) is native to Central America, the West Indies, and most of tropical South America. Its brilliant orange-red bracts are clearly fashioned to catch and store water at their bases. In a natural rain forest setting, these "mini pools", clinging to tree trunks high above the ground, harbor tiny ecosystems of pond organisms such as insect larvae, microbes, and tadpoles.

(BOTTOM RIGHT) *Guzmania sanguinea*, native to Costa Rica, Colombia and Ecuador, is popular these days in hotel lobbies. It is characterized by a broad, flattish rosette of orange- or red-tinged leaves which curl slightly downwards. When flowering, the colors intensify. As you might suspect, *Guzmania's* dramatic effect derives from its modified leaves; look for small yellow flowers when the colorful leaf-spike matures.

✚ CAPE HONEYSUCKLE
(*Tecoma capensis*) Bignoniaceae

Named for the Cape of Good Hope, South Africa, this subtopic shrub has become more popular during the last few years. Look for it at resorts and shopping centers. Upright clusters of narrow, funnel-like, orange-scarlet flowers projecting up from shining green, toothed, compound leaves (7-9 leaflets), are characteristic. Due to lack of appropriate pollinators, fruit does not develop in Hawai'i, thus propagation is by cuttings.

✚ CROSSANDRA
(*Crossandra infundibuliformis*) Acanthaceae

At first glance, crossandra's rich apricot "half-flowers" resemble those of beach *naupaka* (*Scaevola taccada,* see "White" chapter). But no—the flower has evidently adapted to a similar method of pollination, whereby insects are guided into the "cutopen" petals. Crossandra, originally from India, has become more readily available in the 1990s, and thrives well into the upcountry area. Give it shade from wind and it will bloom most of the year. The unusual name means "fringed anthers".

✚ CUP-AND-SAUCER PLANT
(*Holmskioldia sanguinea*) Verbenaceae

The unusual flower shape of this sprawling shrub, native to the Himalayan area, has also inspired names such as Chinese hat plant and parasol flower. The "saucer" is simply a united calyx, while the "cup"—perhaps more like a curved champagne glass?—is a slender, tubular flower with 5 lobes and even longer stamens. The flowers are a little delicate and drop easily, but the calyx is long lasting. Cup-and-saucer plant produces an abundance of flowers to around 4000 feet, and blooms in both sun and shade.

✚ DAY LILY
(*Hemerocallis*) Liliaceae

Ephemeral, like hibiscus blossoms, day lilies last only a single day. This characteristic is reflected in their scientific name, "day beauty". Of Asian origin, many single and double hybrids have been developed with yellow, orange, and reddish hues. The most common day lilies in Hawai'i are the single and double orange. Try sauteeing buds in butter with cashew nuts, or dried in Chinese dishes. The Hawaiian name is *lilia pala'ai* ("pumpkin lilies").

✚ HELICONIAS
(*Heliconia*) Heliconiaceae

Heliconias, the ravishingly elegant and boldly geometrical flowerheads that bespeak the continental tropics, are today the basis of a successful floral trade in Hawai'i. Humid lowlands on Maui (Haiku, Nahiku, Hana) and the windward Big Island are world centers for their commercial production. Visit a Buddhist temple, the library, or a resort—an arrangement with heliconias is bound to greet you.

Heliconias, honoring Mt. Helicon (home of the ancient Greek gods), superficially resemble banana plants. They are easily distinguished by their *absence of banana-like fruit*, and the fact that their *leaves arise directly from ground level rather than from an erect trunk*. See also other chapters.

(ABOVE) *H. psittacorum* x *spathocircinata* cv. 'Golden Torch', a named hybrid called "Parrot Heliconia" in Hawai'i and "Golden Torch" in Florida. Its color is an unadorned, rich orange.

(BELOW RIGHT) Orange latispatha (*H. latispatha*), tall and rambling, is found in large estates and is naturalized in some state parks **(BELOW LEFT)**. It is a tall, medium-sized heliconia, recognized by its pale orange, thin "boats" (bracts), the lowest of which bears a small leaf blade at its tip. This is definitely not one for the home garden! Its natural home is in Central and northern South America, where it commonly colonizes roadsides, streambeds, and light gaps in the forest. Note its short, straight flowers, an adaptation to pollination by short-billed—rather than curve-billed—hummingbirds.

(ABOVE) Parrot's beak heliconia (*H. psittacorum*) cultivars are all short and dainty. Popular in island gardens, they bloom all-year and tolerate drier air and soils better than most other heliconias.

(RIGHT) Jacquinii lobster claw (*H. bihai* cv. 'Jacquinii'), about 18 inches long, typifies a heliconia flowerhead. The red and yellow "floral boats" are actually highly modified leaves, out of which tiny, curved flowers emerge.

(LEFT) The curious, fuzzy orange hanging heliconia (*H. mutisiana*), has both its snaky axis and orange-red bracts clothed in downy hairs. Native to Colombia, it grows ten feet tall. It arrived on Maui in the 1980s and is best seen in the Hana area, for example, Helani Gardens. Each floret matures into a turquoise pearl fit for a queen's necklace. It is related to fuzzy yellow heliconia (*H. xanthovillosa*).

■ HIBISCUS

✚ (*Hibiscus*) Malvaceae

Everyone knows hibiscus these days. Decades of horticultural expertise has resulted in ever-increasing numbers of vivid hybrids, many of which are available on the mainland as summer bloomers. Hibiscus shrubs favor full sun. Frequent pruning assures a continued harvest, maximizing flower production during the summer months. **(TOP)** Most orange hibiscus are hybrids, which are usually too large to tuck behind one's ear, but at least they last for two days. **(BOTTOM)** This dazzling St. John's hibiscus is no hybrid, but one of Hawai'i's native specialties. It is one of several *koki'o 'ula* (*H. kokio st. johnianus*), found only in dry forests in northwestern Kauai from about 500 to 3500 feet. The name commemorates one of Hawai'i's most loved and long-lived botanists, Harold St. John, who passed away in the 1980s.

■ 'IE'IE, A FOREST VINE

(*Freycinetia arborea*) Pandanaceae

'Ie'ie, a native spiny leaved vine, is cousin to pandanus (*hala*). Its woody, climbing stems flourish in Hawai'i's moist montane forests (1000—4000 feet), especially on the Big Island above Kona. In Hawaiian mythology, this sword-leaved vine embodied the eternal spirit of a beautiful maiden, *Lau-ka-'ie'ie* ("leaf-of-the-*'ie'ie*"). **(BELOW)** *'Ie'ie's* (pron. "eeah-eeah") clustered flower spikes, several inches long, nestle within a protective tuft of apricot bracts (modified leaves). After pollination by birds and bats, a thick cylinder of orange fruit develops **(ABOVE)**.

✚ IMPATIENS

(*Impatiens walleriana*) Balsaminaceae

Impatiens grow extremely well along moist roadsides and trails, in gardens, and in hotel lobbies. This salmon-colored variety was photographed in Iao Valley State Park, Maui. On the windward coast they can be propagated simply by cutting off a six-inch cutting and planting it in the ground, without rooting in water first. To prevent "legginess", nip out the growing shoots to encourage the growth of side branches and periodically cut back.

✚ KAFFIR LILY

(*Clivia miniata*) Amaryllidaceae

Also called Natal lily, disclosing its South African origin, this striking orange lily, with strap-like leaves, grows well in the lowlands and upcountry. It usually flowers during summer, preferring shady, well-watered locations. During the non-flowering season, instead of dying back, it continues to produce more of its broad, evergreen, strap-like leaves.

▼ *KOU*, A PACIFIC TREE
(*Cordia subcordata*) Boraginaceae

Kou (rhymes with "throw") is a graceful, widespread Pacific hardwood (30-40 feet tall). It is now believed to be an ancient Polynesian introduction to Hawai'i, after which it became a favorite shade tree until an introduced moth exterminated almost every one. Recently, it has been reintroduced for landscaping (for example, Wailuku, Kahului) and trees are healthy. Similar to *kou-haole* (**BELOW**), it is easily distinguished by *glossy, smooth leaves* and *brown (dry) fruits* rather than *sandpapery leaves* and *fleshy, white fruits. Kou's* crepy flowers make unusual leis, its leaves and fruits are medicinal, and its reddish-grained wood has long been esteemed for furniture, bowls, canoes, etc., a major reason for its historial decline elsewhere in the Pacific.

✚ *KOU-HAOLE* OR FOREIGN *KOU*
(*Cordia sebestena*) Boraginaceae

Shading cars at Kaahumanu Shopping Center and flanking Alamaha Street, Kahului are trees with stiff, sandpapery leaves, brilliant orange, crepy, tubular flowers, and white fruits. Native to the West Indies, this cheery ornamental almost qualifies as a native, since for many years it replaced a very similar tree indigenous to Pacific islands, the *kou* (**ABOVE**). Now Maui has both. Note the flower's "nectar guides," radiating like spokes of a wheel, indicating the nectar source to insects.

✚ LANTANA
(*Lantana camara*) Verbenaceae

(**RIGHT**) The familiar gay-hued lantana, an escaped ornamental throughout Hawai'i, is a serious agricultural pest: numerous insect parasites have been introduced to control it. Arid rangelands are commonly infested with lantana. Not only do its prickly branches and oily, pungent odor deter cattle, but pasturelands are further destroyed as mynah birds disseminate its prolific blackberry-like fruits. Children have died from eating its poisonous fruits, which are also poisonous to cattle.

✚ MILKWEED
(*Asclepias curassavica*) Asclepediaceae

Also known as butterfly weed, bloodflower in English, and *laulele* or *lauhele* in Hawaiian, this perennial weed (superficially resembling lantana, above) is found in dry low elevation habitats. Although introduced early to Hawai'i (mid-1840s), it is generally not considered a pest. This particular milkweed is native to Florida, the West Indies, and South America. Milkweed flowers are elaborately constructed so that they are unable to fertilize themselves. The flower is odd: its petals and sepals are so downcurved they are almost inside-out and the upper corona (which resembles a diamond ring setting), is composed of fancy nectar "hoods". An insect alights on the corona and as its feet slip into the slits between the hoods, pollen adheres to them.

✚ MONTBRETIA
{Crocosmia x *crocosmiiflora* (= *Tritonia crocosmiiflora*)} Iridaceae

A striking orange, lily-like flower, with stiff, upright leaves, montbretia belongs in the iris family, closely related to gladioli. A hybrid of two African species (*Crocosmia aurea* and *C. pottsii*), it grows wild and in gardens. Rarely are there one or two—usually they appear as a dense ground cover, creeping and spreading by means of sturdy underground corms and stolons.

✚ ORCHIDS
Orchidaceae

Perhaps more than any other flowers, orchids epitomize the exuberance and diversity of the tropics. Many people assume that Hawai'i's rain forests are dripping with exotic orchids. However, our planet's geography and the peculiar survival requirements of orchids—symbiosis with specialized fungi and insect pollinators—prevented this from happening. Hawai'i, over 3000 miles from plant source areas in Central America and 4000-6000 miles from Australasia, was never provided opportunities to evolve a significant native orchid flora, despite the fact that orchid seeds are feather-light and known to be transported high in trans-Pacific jet streams.

(RIGHT) A medium-sized hybrid orchid, *Epicattleya* cv. 'Fireball Palolo', exhibits genetic traits inherited from both parents, *Cattleya* and *Epidendrum:* a frilly *Cattleya*-like lip and simple elliptical petals and sepals which recall *Epidendrum.*

(BELOW LEFT) Hybrids of the major orchid of commerce, *Cattleya,* sporting 4 to 5-inch flowers, are commonly seen in Hawai'i, although orange flowers are uncommon. Some are fragrant, others are not. It's always worthwhile to check.

(BELOW RIGHT) In upcountry Maui (1000—3000 feet), an orange flowered epidendrum (*Epidendrum* x *obrienianum*) called scarlet orchid, may be found on rocky slopes or in gardens. Its small, fringed flowers, usually orange, also come in red or mauve (see other chapters). Terrestrial, this epidendrum grows upright (to four feet), developing dense masses of stems as it matures. This is one of the earliest orchid introductions into Hawai'i (1940), and is easy to grow. An alternate name, "baby orchid," refers to its tendency to develop tiny plantlets along its older stems. These transplant easily.

✚ PRICKLY PEAR

(*Opuntia ficus-indica*) Cactaceae

Drivers in upcountry Maui and other aridlands in Hawai'i cannot fail to notice the large area of pasturelands occupied by prickly pear (*panini*). Orange-red summer flowers quickly ripen into blood-red, egg-shaped fruits, perfect for jam and juice. The yellow fruits are also tasty ... but, potential foragers, BEWARE. However tempted you are to sample these extra-sweet fruits "in the field," *do not even think of it.* Not only do they have nasty visible spines (to one inch long), worse by far is their invisible overcoat of microscopic spines which lodge in your fingers, lips, gums, and tongue, irritating you for *days.* Choose a calm day, cover yourself, use tongs, scrub the fruit under a hose, *then* enjoy them! Treat the young, fleshy "joints" similarly, and cook as a green vegetable, Mexican-style.

✚ PINCUSHION PROTEAS
(*Leucospermum*) Proteaceae

These curious botanical novelties, originally from South Africa, mushroomed in the 1980s into a lucrative floral enterprise. Primarily grown above 3000 feet on the cool slopes of Haleakala, (Maui) and Kohalas (Big Island), pincushions (locally dubbed "sunbursts") exhibit a dazzling array of shapes and sizes: bursts of orange fireworks, fleecy "pinecones", looped corncobs, golden spiky acorns, etc. Variations on a basic orange hue include pink-apricot, and fiery orange-red. **(RIGHT)** Maturing in ever-increasing splendor, a "rocket" pincushion (*L. reflexum*) unfurls its "pins" outwards and downwards, doubling the flower size.

(CENTER TWO) Richly glowing "sunburst" pincushions (*L. cordifolium*), the first species grown in quantity outside Africa, are the best known of all proteas. On Maui they are abundant, cheap, and available all-year, especially in winter and spring. Farms are in Kula and near Waimea, Big Island. Pincushion leis are beautiful, bulky, and often presented to civic dignitaries.

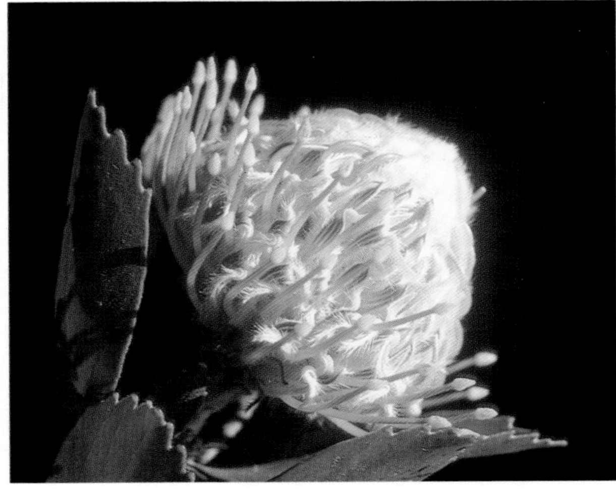

(BOTTOM RIGHT) Orange, red, and gold streaks nestle within and burst forth from a dome of white fluffiness: Veldfire (*L.* cv. 'Veldfire').

✚ RED BIGNONIA
(*Distictis buccinatorius*) Bignoniaceae

Although this lovely orange, scarlet-and-yellow viney shrub grew in our yard for years, I didn't know its name and no-one else seemed to either. I discovered that it is Mexican, with various appelations including *Bignonia cherere, Pithecococtenium buccinatorium, Phaedranthus buccinatorius,* and scarlet trumpet vine. Red bignonia is a bright-hued addition to anyone's garden, requiring little water and care and bursting forth with masses of blossoms all-year. Its flowers also attract Japanese White-eyes and other birds.

✚ SHOWER, RAINBOW
(*Cassia* x *nealiae*) Fabaceae (Leguminosae)

Sometimes it is difficult to decide the "correct" color of rainbow shower's flowers. There is no correct color! Every tree bears different complements of lemon, darker yellows, pink, rose, bronze, and so on. Several named cultivars range from cv. 'Madge Tennet'—the deepest of the "pinks"—to 'Queen's Hospital"—pale, pale yellow to whitish—and several named shades of yellow. When not blooming, rainbow showers can be recognized by their dark green, oval leaves. In contrast, golden showers bear longer, light green leaves, while pink-and-white shower leaves are oval and small. See also "Yellow" chapter.

■ *WILIWILI* OR HAWAIIAN CORAL TREE
(*Erythrina sandwicensis*) Papilionaceae (Leguminosae)

Wiliwili Dry Forest was once widespread in arid habitats throughout Hawai'i. Only remnant patches of this endemic tree persist. A good area is along the rough "back" road between Ulupalakua and Hana, around 1500 feet elevation. *Wiliwili* is a type of coral tree (see "Red" chapter), but its flower clusters range from chartreuse to red-orange, commonly apricot.

In old Hawai'i, the light wood of this swollen-trunked, baobab-like tree, was fashioned into surfboards, fishnet floats, and canoe outriggers. It is sometimes carved into imitation whale-tooth necklaces (*lei palaoa*). *Wiliwili* means "twisting", referring to the twisting action of seedpods as they liberate their seeds.

A splendid tropical America heliconia, "Sexy pink" (*Heliconia chartacea* cv. 'Sexy Pink'), was named by Maui's foremost heliconia grower of the 1980s, Ali'i Chang.

PINK FLOWERS

PINK, A UBIQUITOUS tropical flower color, is well represented on Maui. Shades include rosy-blushed white, pinkish-mauve, flesh-colored, cerise, salmon, and pinkish-red, as well as pale, "regular," and gaudy pinks. Often the same species may show considerable variation in hue. Bees and butterflies, owing to their ultraviolet vision, frequently see contrasting patterns on pink flowers that are invisible to us; for example, central discs and darker radiating spokes typically focus into a hidden nectar source. This chapter includes some abundant tropical flowers as well as Hawaiian specialties bearing the dubious distinctions of being some of the world's rarest plants.

(RIGHT) An unusual lei featuring Mexican creeper (*Antigonon leptopus*), curling to the right; smartweed (*Polygonum capitatum*), the little pale pink balls; dark red China asters (*Callistephus chinensis*); a maroon *'ohi'a lehua* (*Metrosideros polymorpha*), rosebuds and maidenhair ferns. Complex leis with intertwining flowers and foliage are called *haku,* or if everything is sewn to a backing, *humupapa.*

■ 'AKALA, AN HAWAIIAN RASPBERRY
(*Rubus hawaiiensis*) Rosaceae

Fruits and seeds on islands often exhibit giantism—a tendency to enlarge more than their continental counterparts. *'Akala* (pron. "ah-kah-lah") is a prime example, since it bears deep rose fruits one inch in diameter. They were formerly a source of pink dye for tapa (barkcloth). Although they taste less sweet than our familiar cultivated raspberry, they nonetheless make good jam. *'Akala* is a rambling shrub, typically found in gullies or sunny clearings within moist montane forests (Olinda, Kula, Haiku). *'Akala's* attractive pink blossoms resemble a single wild rose ('akala means "pink"). *Kala* also means "spiny". Hawai'i has a native spiny poppy, *pua kala* (lit. "spiny flower", see "White" chapter) and coastally, sargassum seaweed, *limu kala* (lit. "spiny seaweed").

✚ ANNATTO OR LIPSTICK PLANT
(*Bixa orellana*) Bixaceae

We are so accustomed to recognizing annatto by its red or brown spiky seedpods **(RIGHT & BELOW)**, it sometimes goes undetected if only bearing flowers. The rose-like flowers are actually beautiful, with 5 pink petals and numerous pink stamens. Look for it in gardens, particularly in Makawao and Kahului. Annatto's numerous red seeds remain today the chief source of dye for cheese, margarine, and butter.

✚ ANTHURIUMS

(Anthurium andraeanum) Araceae

(RIGHT) At home in the muggy rainforests of tropical America, all anthuriums flourish in high humidity and warm, year-round temperatures. In Hawai'i these natural conditions are best met for commercial purposes on the Big Island. On other islands, anthuriums are primarily grown by homeowners: outdoors in humid areas or in greenhouses. Indoors, they respond well to frequent misting. Anthuriums are seldom used in resort landscaping because of wind, salt and dryness. **(BELOW LEFT)** Pink anthuriums are commonly utilized in island-wide flower arrangements. Here pink and *obake* anthuriums accent lilies (*Lilium*), pink ginger, and caribaea 'Cream' heliconias. See also other chapters.

✚ AZALEA

(*Rhododendron indicum, R. macrosepalum* and *hybrids*) Ericaceae

Azaleas, more at home in the cooler climates of the world, are surprisingly well represented in Hawai'i. Most are hybrids from Japan, and many beautify Japanese lowland gardens (Wailuku, Kahului) but are also popular with residents upcountry. The higher in elevation, the more prolific azaleas become. However, in Hawai'i they never produce the solid masses of color seen in the temperate zones.

Cameron Kepler

✚ BANANA POKA

(*Passiflora mollissima*) Passifloraceae

In 1926 banana poka, a beautiful Andean passionfruit vine appeared on the Big Island. Purpose: to hide an outhouse. After this minor success, its vigorous growth proceeded to blanket and strangle many square miles of native forest on Kauai and the Big Island **(LEFT)**. An easy place to see it is at Kokee, Kauai.

In Hawai'i, banana poka is now a serious pest: it has no natural controls. Feral mammals gobble its tasty fruit, spreading seeds widely. NEVER PLANT IT ANYWHERE. **(LEFT)** Dangling blossom, 3 to 4 inches long, with large, pulpy fruit. **(BELOW RIGHT)** "Face" view of flower, 2-3 inches in diameter.

Cameron Kepler

■ BEGONIAS

✚ (*Begonia, Hillebrandia*) Begoniaceae

Many species and hybrids of begonias are grown on Maui, with new varieties arriving constantly, for example wax begonias (*Begonia semperflorens*) and taller hybrids **(RIGHT & BOTTOM RIGHT)**. With shade, they bloom constantly, but upcountry they need protection from temperatures below 50° during winter.

(BELOW) Hawai'i has its own lovely native begonia (*Hillebrandia sandwicensis*), found in shady gullies on East Maui (1800-5500 feet). Like all begonias, the 4-petaled flowers are crisply succulent, with soft, lopsided leaves. Its two flower types (male and female) droop together in graceful clusters. The colored flower parts are not petals but bracts and sepals. Note the numerous yellow stamens in the male flowers—an inspiration for *pua-maka-nui* ("large-eyed flower")? The plant honors William Hillebrand, a Honolulu physician who published "The Flora of the Hawaiian Islands", Hawai'i's first botanical treatise, in 1888.

✚ BRAZILIAN BOWER PLANT

(*Adhatoda cydoniifolia,* = *Megakepasma erythrochlamys*) Acanthaceae

This showy tropical is not common, but can easily be seen in the Wahiawa Botanical Garden (Oahu) and growing along the roadside in the Nahiku section of the Hana Highway (Maui). The Brazilian bower plant (also Brazilian red-cloak, red justicia) grows to about 15 feet high, bears broad, oval leaves and bright pink bracts with two-lipped white flowers. It is a popular garden ornamental in tropical Central and South America, where it is native.

✚ BROMELIAD

(*Aechmea fasciata*) Bromeliaceae

Candy-pink, spiky "flowers" arising from a basal rosette of horizontally striped leaves, characterises *Aechmea fasciata,* which has no common name. Close inspection of the pink "petals" reveals that they are our old friends, bracts—modified leaves. Note the tiny purplish-blue true flowers peeking out from behind their spirally enfolding bracts. *Aechmea,* perhaps the most widely known ornamental bromeliad, graces hotel lobbies and private gardens. These days, you might even find it in K-Mart! Originally from Brazil, *Aechmea* is long-lasting and hardy indoors. Note a strong similarity to the flowers of commercial pineapple (this chapter).

✚ CARNATIONS

(*Dianthus caryophyllus*) Caryophyllaceae

Pink, red, white, mauve, yellow, and "peppermint" **(LEFT)** carnations, dot Haleakala's mid-level slopes from 3000 to 4000 feet elevation. Maui grows over 30 million carnations annually. Some farms are tucked away from Maui's major roads, but others are visible along routes 377 and 378 **(RIGHT).** If you wish to make your own leis (each flower costs only a few cents), consult the phonebook's yellow pages under Flowers, Wholesale. However, remember that because the farms are not in the sunny lowlands, extended periods of cloudy and rainy weather upcountry inhibit flower production. In some winter months farmers barely meet their commercial obligations.

✚ FUCHSIAS
(*Fuchsia*) Onagraceae

These cool-loving, shrubby ornamentals flourish in mid-elevation residential areas statewide. The common name derives from Leonhart Fuchs (1501-1566), a German physician and herbalist. Although there are approximately 100 natural species—originally from Mexico to Patagonia— most of the hundreds of cultivars have been developed from three species. The fuchsia's pendent blossoms are frequently two-toned (pink, purple, or red). Fuschias can be combined with agapanthus or other smallish flowers to make attractive leis. Bushes bloom prolifically if heavily pruned after the flowering stems become "leggy".

▼ GINGERS
✚ Zingiberaceae

Although the word *ginger* conjures up mental images of Oriental food, edible ginger is only one of around 1400 species in this highly diverse family. Several beautiful ornamental gingers, intimately associated with island beautification as naturalized escapees or as "staples" in *kama'aina* (old-timers') gardens, are assumed to be native, even by residents. Everywhere they occur, the aromatic flowers, creeping stems, or seeds of gingers are valued culturally. See other chapters for different colored gingers.

(ABOVE) Shampoo, wild ginger or 'awapuhi (*Zingiber zerumbet*) was brought by early Polynesians and used for perfuming tapa (barkcloth) and wrapping food. Try shampooing your hair with the fragrant, sudsy juice squeezed from its flowerheads. **(LEFT)** Resembling porcelain shells, clusters of shell ginger flowers (*Alpinia zerumbet*) grow in gardens and wild along windward (moist) roadsides.

(LEFT) Pink ginger, *Alpinia purpurata* cv. 'Eileen McDonald' arrived in Hawai'i from Tahiti in 1973, and is now common. Look around resorts and office complexes. It is also pictured on page 2. Pink ginger is derived from red ginger (see "Red" chapter) and is named after a Mauian. **(BELOW)** The majestic torch ginger (*Etlingera elatior*), native to Mauritius, Indian Ocean, is long-established in Hawai'i. Look for its 20-foot-high leaves in arboretae throughout the state, and occasionally along windward roadsides, especially on Kauai and Maui. One of its former scientific names (*Phaeomeria magnifica*), still found in books, means "magnificent, pure light".

✚ HELICONIAS
(*Heliconia*) Heliconiaceae

"Tropicals"—a word coined in the 1980s by Maui's Ali'i Chang—incorporates three major groups of plants: heliconias, gingers, and tropical foliage. The dramatic color and form of a heliconia bloom is not due to its flowers (which are small, inconspicuous, and rot easily), but rather to bracts, the highly modified leaves that provide dazzling color in many unrelated families (bougainvillea, gingers, bromeliads, proteas). Each overlapping "floral boat" is a bract which houses the tiny true flowers. These are easily seen in 'Sexy Pink', *Heliconia chartacea* cv. 'Sexy Pink' **(BELOW & CHAPTER HEADING)**, a stunning heliconia bedecked in dark rose, erotic pink, and translucent apple-green.

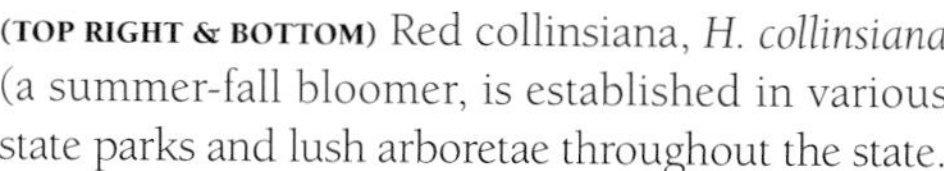

(TOP RIGHT & BOTTOM) Red collinsiana, *H. collinsiana* (a summer-fall bloomer, is established in various state parks and lush arboretae throughout the state.

(ABOVE) 'New pink' (*H. colgantea*), only discovered in the 1970s, is similar to red collinsiana but has more orange flowers, a steeper bract angle, and lacks the white powdery bract coating. See also "Pink", "Yellow", and "Orange" chapters.

✚ HIBISCUS
■ (*Hibiscus*) Malvaceae

Who can resist the flamboyant charm of a big hibiscus, gaudily arrayed with flaring petals, ruby throat and an elegant staminal tube? This central protrusion unashamedly proclaims its hermaphroditic nature by thrusting out—inches away from the flower body—tiny stalked spheres of golden pollen (male) and velvety, knobbed stigmas (female). Hundred of hibiscus hybrids now flood nurseries throughout the country. Pictured are single and double varieties. In tropical American forests, conspicuous blooms like these regularly attract hummingbirds, which guzzle their sweet nectar then automatically cross-pollinate as feather-borne pollen from one flower brushes onto the stigma of another.

(CENTER RIGHT) A rare Hawaiian native, *'akiohala* (*H. furcellatus*), formerly flourished throughout the islands. Lavender-pink flowers with a dark purple "eye" and sandpapery, heart-shaped leaves, characterize this erect shrub (to 12 feet). Little more than a century ago its upright stalks were so plentiful, untold thousands were uprooted for house thatching. Nutritionally, *'akiohala* was a valued source of leafy greens, especially for *hapai* (pregnant) women, since is was considered "to make whole and firm the body of the child".

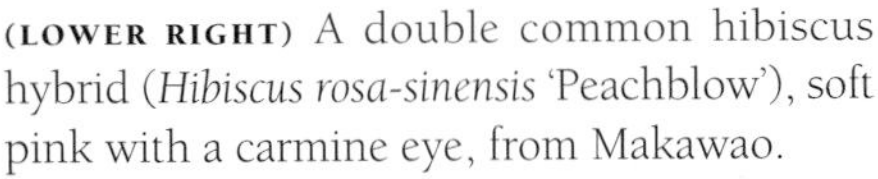

(LOWER RIGHT) A double common hibiscus hybrid (*Hibiscus rosa-sinensis* 'Peachblow'), soft pink with a carmine eye, from Makawao.

✚ IMPATIENS
(*Impatiens walleriana*) Balsaminaceae

Cheery touch-me-nots—pink, apricot, red, and white—perk up lush windward roadsides, shady garden nooks, and hotel lobbies. Little children enjoy pinching the plump seedpods to pop out their seeds—in so doing, they prematurely mimic the plant's natural seed dispersal mechanism. These succulent herbs, also called jewel weeds, busy lizzie, or sultan's flower, originated in Zanzibar, east Africa. Recently a plethora of hybrids have appeared on the commercial market. My favorite name is the Puerto Rican version, *miramelinda:* "look at me, I'm pretty!"

✚ INKBERRY OR SHOEBUTTON ARDISIA
(*Ardisia elliptica*)Myrsinaceae

Inkberry is Maui's local name, dubbed by people in the Hana area for this pestiferous tree which bear masses of inky-purple berries. During the 1970s it was primarily restricted to the Waianapanapa/Hana airport area, but is spreading widely throughout the entire windward coast. Its star-shaped, pinkish-lavender flowers are quite attractive, as are its fresh pink leaves, but unfortunately its energetic nature and great capacity for seedling germination is crowing out great swaths of Maui's windward forests.

✚ JATROPHA, ROSE-FLOWERED
(*Jatropha integerrima*) Euphorbiaceae

Gay, rose-colored jatropha blossoms enliven shopping malls, residences and resorts throughout Hawai'i. Although tolerant of aridity, they prosper best in richly humid areas such as Hana and Kipahulu. Their five-petalled flowers resemble oleander, but are almost odorless, and their leaves are wide and long-stemmed rather than narrow and short-stemmed. Jatropha, from Cuba, is a "pollen flower": observe how the stamens, copiously loaded with pollen, advertize their golden brilliance by being raised above the general flower level. What bee could resist such a meal?

■ *KAPANA*, A CLOUD FOREST MINT
(*Phyllostegia ambigua*) Labiatae

Closely related to garden mint, but odorless, *kapana* (pron. "kah-pah-nah") is one of Maui's botanical specialties: its entire range lies between 3000 and 6000 feet on Haleakala. Widely scattered individuals spring up from mossy rainforest carpets or border soggy alpine bogs. *Kapana's* mint-like flowerheads, composed of a dozen or more lipped flowers, are always rain-soaked. How translucent are those blushing pinks! And how cleverly the "hood" protects the most important part of the flower—its reproductive parts. In order to safeguard Maui's delicate wet ecosystems from feral mammals, huge areas of montane forest have been fenced by the National Park and State of Hawai'i.

■ *KOLI'I*, A "CANDELABRA" LOBELIA (*Trematolobelia macrostachys*) Campanulaceae

One of Hawai'i's most spectacular native plants is the koli'i which, like kapana, is adapted to high montane bogs and rainforest edges. An unbranched shrub reaching 12 feet tall, its symmetrical, candelabra-like form is a delight to the eye: five to twenty arms **(RIGHT)**, each crowded with slender cerise flowers **(BELOW)**, radiate like wheel spokes from the central axis. Lucky observers may spot Hawaiian honeycreepers visiting koli'i's floral bounty to suck nectar. Its closest relatives reside on other South Pacific high islands (Marquesas, Tahiti, Cook Islands).

Haleakala National Park/Betsy Gagné

✚ *LOKELANI*, MAUI'S ISLAND FLOWER (*Rosa* x *damascena*) Rosaceae

Maui's official island flower, the double-flowered *lokelani* (pron. "low-kay-lah-nee") rose, is suffused with a delicious, old-fashioned fragrance **(RIGHT)**. Unfortunately its bushes are so uncommon that a relatively insipid, modern rose is normally substituted in island celebrations. Occasionally, however, one encounters leis of the authentic flower ... what a pleasure! Native to Eurasia, *lokelani* was introduced to Hawai'i in the 1820s by missionaries. Outside of Hawai'i, it is known as Damask rose, a "lovely old Rose of long ago ... with rich velvet texture and grand fragrance". This species is the source of attar of Roses and rose water.

(LEFT) A pretty Maui teenager, Lauralei Royster, wears a *lokelani* headlei. (RIGHT) Island *kama'aina* Eda Kinnear celebrates Maui's official flower with bouquets and a hand-crafted quilt. Hopefully, the future may provide opportunities for more Mauians to grow *lokelani*. Look for it at the Keanae Arboretum and in Waihee and Wailuku.

✚ MANUKA OR AUSTRALIAN TEA TREE
(*Leptospermum scoparium*) Myrtaceae

Long popular in southern gardens, this heath-like shrub is very drought-tolerant, can withstand cold temperatures, and is most amenable to producing a variety of attractive hybrids. On Maui, Big Island, and Kauai, manuka is primarily found in cool upcountry gardens, where flower colors range from pale pink to dark rose, single or double. The plant may be a low bush, medium shrub or small tree. Native to Australia and New Zealand, leaf infusions were made into "tea" by early colonists. Today its therapeutic properties are available in health food stores in the form of oils and toothpastes.

✚ MEDINILLA OR ROSE GRAPE
(*Medinilla magnifica*) Melastomataceae

Although an unusual and beautiful ornamental, medinilla has been declared a noxious plant by the State of Hawai'i (see Environmental Alert). During the last 10 years, it has become increasingly available in Maui nurseries, and is particularly prevalent in the Nahiku-Hana area. Every one of its copious berries is potentially several new plants. If it were from any family other than Melastomataceae, perhaps it would not matter...but all melastomes spell disaster for Hawaiian rainforests. Please do not grow it.

✚ MEXICAN CREEPER

(*Antigonon leptopus*) Polygonaceae

This eye-catching Mexican vine with copious strings of bright pink flowers and climbing tendrils, is not especially common in Hawai'i. It mantles chain-link fences, rock walls, or other appropriate climbing foundations. Other names are hearts-on-a-chain, chain of love, and coral vine. It is surprising that this endearing liana is not a Valentine's Day symbol, since its buds, flowers, and leaves are all heart-shaped. Perhaps the larger anthuriums state the "love theme" better, or perhaps Mexican creeper are too reminiscent of bleeding hearts (*Dicentra*)!

✚ MONKEYPOD

{*Samanea (= Albizia) saman*} Mimosaceae

This familiar spreading tree, with its pink, powder-puff blossoms, lacy foliage, and rounded crowns, is abundant throughout Hawai'i's lowlands. Although most of its pleasingly grained wooden bowls and other carvings are plentiful in curio shops, they are mostly imported from the Philippines. Monkeypod, however, originally haiils from tropical America, where it is home to a staggering variety of birds and insects. The name "monkeypod" derives, not from monkeys, but from its former scientific name, *Pithecellobium,* "monkey's ear-rings". This alludes to its bulbous, pea-like seedpods, 4 to 8 inches long—presumably earrings fit for a monkey!

✚ MOSS (GARDEN) VERBENA

(*Verbena* x *hybrida*) Verbenaceae

This gaily colored ground cover exists in a complex array of cultivars and hybrids derived from several species, all South American. Its creeping stems sprawl rapidly, simultaneously budding off masses of small tubular blossoms, making it an excellent ground cover. Although an annual on the mainland, in Hawai'i's equable climate moss verbena grows perennially both in the lowlands and upcountry. See also "Purple" chapter.

▼ MOUNTAIN APPLE

(*Syzygium malaccense*) Myrtaceae

(CENTER RIGHT) As you drive windward highways or hike in wet forests during May and June, hundreds of broad-leaved trees suddenly burst into vibrant life as showy sprays of rose pink "shaving brushes" appear. An early Polynesian introduction (originating in Malaysia), mountain apples are now extensively naturalized throughout Hawai'i's moist lowlands from 600 to 1000 feet elevation. The people of old named this tall tree (to 80 feet) *'ohi'a 'ai,* recognizing its resemblance to Hawai'i's ubiquitous forest tree, *'ohi'a lehua.* (BELOW) Mountain apples are sweet, with a smoother-than-pear texture.

✚ MUSSAENDA

(*Mussaenda erythrophylla 'Rosea'*) Rubiaceae

During the last 15 years, mussaendas (pron. "mouss-aye-enda") have increased in popularity in Hawai'i and other Pacific islands. Masses of eye-catching, pink, droopy "leaves" characterize this rounded bush (to 10 feet tall)—but wait, there are green leaves also! Is this like bougainvillea, where colored, leaf-like bracts (modified leaves) surround the central flowers? It's similar ... the pink "leaves", instead of being modified leaves or petals, are sepals (on most other flowers, these are small, green, and hidden beneath the petals). Thus each pink leafy cluster is the bottom part of a flower, while the golden "star" (5 united petals), is the top part. Originating in tropical Africa (where it is called Red Flag Bush), mussaenda now sports red-, white-, and pink-calyxed varieties. "Mussaenda" is a Sri Lankan name.

■'*OHA-WAI* AND *HAHA*, TREE LOBELIAS
■(*Clermontia* and *Cyanea*) Campanulaceae

Hawai'i's native lobelias and their attendant pollinators, Hawaiian honeycreepers, are showcases of co-evolution. From only five original introductions, more than 110 species have evolved. A noted, early botanist, Hillebrand, called them the "peculiar pride of our flora". *Haha* and *'oha-wai* are general names that apply to many species throughout Hawai'i. Sadly, feral pigs eat lobelias voraciously, but Hawai'i's forests still nurture several beautiful species, although populations are small.

(ABOVE) *Cyanea angustifolia,* reasonably common in parts of the Koolau Range (Oahu), is much rarer on Maui, Molokai and Lanai. In full bloom, hundreds of curved blossoms create a stunning, roseate globe. **(RIGHT)** Another *Cyanea angustifolia,* this time from Castle Trail, eastern Koolau Range, Oahu. This is one of Hawai'i's most spectacular native lobelias, its branched shrub reaching 15 feet tall. Peak blooming is in September.

Cameron Kepler

(LEFT) Warty clermontia, an *'oha-wai* (*Clermontia tuberculata*) is a pink, knobby oddity with an extremely restricted geographical range: 5500 to 6000 feet along a few streambeds in East Maui.

■'*OHELO*, AN EDIBLE ALPINE BERRY
(*Vaccinium*) Ericaceae

"*Ohelo,* one of the most common shrubs at high elevations (especially Maui and the Big Island) is related to blueberries and best known for its 1/4-inch, red fruits. Eagerly munched by hikers, it was formerly baked into tasty pies. Hawai'i has three endemic species of *'ohelo.* Commonest is *V. reticulatum*, shown with small, pinkish-yellow flowers (also red-yellow, yellowish-green) and juicy red fruits. *'Ohelo* bears flowers and fruit all-year, with a concentration of flowers in spring and fruit in fall. *'Ohelo's* principal kin hail from north temperate regions. Over millenia, migrating birds have distributed them to, and between, Pacific island mountaintops: Hawai'i, Cook Islands, Tahiti, Samoa, and Fiji.

✚ OLEANDER

(*Nerium oleander*) Apocynaceae

(RIGHT) Oleander needs no introduction to Hawai'i residents and visitors from warmer climes such as California and Florida. It is native to a wide swath of Eurasia from the Cape Verde Islands (off Africa) to Japan. Recently, a wider variety of colors has become available, including a velvety, cerise cultivar **(BELOW)**. Enjoy oleander's almondy spiciness (not all are fragrant), but no matter how desperately you need sticks to roast your hot dogs, or how much your toddler insists on chewing its leaves, remember that it has been used in rat poisons and has rung the death toll for humans too.

✚ ORCHID, PHALAENOPSIS OR MOTH ORCHID

(*Phalaenopsis*) Orchidaceae

Many pink orchids in Hawai'i belong to a genus called *Phalaenopsis* (literally "moth-like orchids"), a group of 40 species native to rainforests of southeast Asia and Australia. Under the artful expertise of horticulturalists, these natural species have burgeoned into a plethora of exquisitely "painted" hybrids, whose gently arching strings of long-lasting, blossoms never fail to please. Since *Phalaenopsis* orchids have no pseudobulbs (fleshy water reserves at their stem bases), they require wind-protection. They are thus usually encountered indoors. Pictured is *Phalaenopsis* cv. 'Pink'.

✚ ORCHID TREES

(*Bauhinia*) Caesalpiniaceae (Leguminosae)

Totally unrelated to true orchids, orchid trees are tropical legumes with long, bean-like seedpods and elegant, curvaceous blossoms which burst open amid a mass of butterfly shaped leaves. The pink bauhinia or St. Thomas tree (*Bauhinia monandra*) has long been planted as a specimen tree in Hawai'i, especially in coastal leeward areas. Aala Street in Honolulu has dozens flanking the road. Peak flowering is in spring. See "Purple" chapter for related species.

■ *PA'INIU*, A RAIN FOREST LILY
(*Astelia menziesiana*) Liliaceae

Pa'iniu is best spotted by *gestalt* ("jizz" or undefinable "look"). This shapely, bromeliad-like, native lily grows either on the ground or as an *epiphyte* (a non-parasitic perching plant which derives its nourishment from air, rainwater, and organic debris on its support substrate). *Pa'iniu* is common in wet forests, preferring elevations from 2000 to 6000 feet. **(RIGHT)** Hikers in the Alakai Swamp (Kauai) or in rainforests such as Waikamoi (Maui), Hawai'i Volcanoes National Park (Big Island) or Kamakou (Molokai) cannot miss its lanky rosettes of narrow silvery leaves adorning branches and tree trunks. **(LEFT)** Hundreds of tiny, flesh-colored lilies crowd together on long, spreading panicles. Male and female flowers occur on separate plants. The people of old incorporated *pa'iniu* flowers and leaves into woven leis.

✚ PINEAPPLE
(*Ananas comosus*) Bromeliaceae

In Hawai'i pineapples, native to Brazil, arrived soon after Captain Cook (1778). This began a long history of commercial production. In Hawai'i today, plantations cover about 40,000 acres, with an average yield of 30-35 tons/acre. At the peak of summer harvest, around 12,000 people are employed in fields, canneries, and offices. Everyone salivates at the mental image of a luscious, field-ripened pineapple but few think about its beautiful, compound flowerhead. Their artistic structure—spipky globes of overlapping pink bracts, from which pinkish-purple flowers emerge—recalls proteas and *Aechmea* bromeliads (this chapter).

✚ PINK TECOMA
(*Tabebuia rosea*) Bignoniaceae

Towns on the dry side of Hawai'i's islands are best for pink tecoma. Since this tall, rather narrow tree is drought resistant, it beautifies shopping centers, office landscaping, parking lots, etc. It blooms early in the spring, dropping its pale, fragile flowers on sidewalks and lawns. In its native homelands (tropical America), where it is called *roble* ("oak"), this beautiful flowering hardwood is usually alive with tanagers and hummingbirds. Recent biomedical research has shown that the bark from several species of *Tabebuia* is helpful in the ongoing fight against cancer.

✚ PLUCHEA, INDIAN
(*Pluchea indica*) Asteraceae (Compositae)

A native of southeast Asia, pluchea (pron. "PLU-keeah") is a scrubby weed (to six feet tall) that for obscure reasons has been introduced into many tropical areas. Also called Indian fleabane, Indian pluchea hardly deserves inclusion in a book entitled *Hawai'i's Floral Splendor,* but it (or a hybrid) is locally common throughout the islands, especially in saline habitats. Its rounded flowerheads, composed of many tiny daisies, are a soiled pinkish-purple color, while the toothed leaves are a nondescript green, usually dry and curled. Needless to say, you won't find pluchea in gardens.

✚ PLUMERIA
(*Plumeria rubra*) Apocynaceae

Intoxicatingly fragrant and richly sentimental, plumerias are all-time favorites. Hawai'i is liberally sprinkled with plumeria trees from the lowlands to about 3300 feet. Botanically, there are seven species, all tropical American, plus numerous hybrids. In the West Indies, wild *P. rubra* ("red plumeria") looks just like the photo here with the richest color. The most deciduous plumerias are *P. rubra,* regardless of flower color (see also "White" and "Yellow" chapters). In natural forest settings, plumerias bear fewer leaves and flowers than in cultivation. Anyone can grow plumerias: simply break off a deer antler-like branch a few feet long, lay it in the shade for a day or two to heal the broken end, then plant it in a deep fertilized hole.

✚ PODRANEA OR PINK TRUMPET VINE

(*Podranea ricasoliana*) Bignoniaceae

The arching stems of this hefty pink-flowering vine may reach 15 feet in length. Naturalized in several areas, it is particularly noticeable in Kula, Maui. Podranea is an odd name, invented by botanists. It is an anagram of *Pandorea,* one of the genera to which it formerly belonged. Its original home is in the Cape of Good Hope area, South Africa, thus it tolerates slight frosts (when it becomes deciduous) and dry soils. Note the flower's attractive rosy veins and darker central area, inviting insects to inspect its floral wares.

✚ PROTEAS

(*Protea, Leucospermum*) Proteaceae

Pink proteas are a diverse group, primarily grown on Maui and in the Kohala region, Big Island. It is impossible to cover them all here (see other chapters and *Proteas in Hawaii* by Kepler & Mau, Mutual Publishing). **(RIGHT & BELOW)** One of the most regal is the king protea, South Africa's national flower. It has been described as "possibly the most spectacular flower in the world". Six to seven inches in diameter, the "king" is composed of a fuzzy-white central mound encircled by green, pink, and rose petal-like bracts (modified leaves). Its earliest illustration (1705) was inscribed "African tree-artichoke", an inspiration for its botanical name, *Protea cynaroides,* "protea like a globe artichoke." A pretty good comparison ... don't you think?

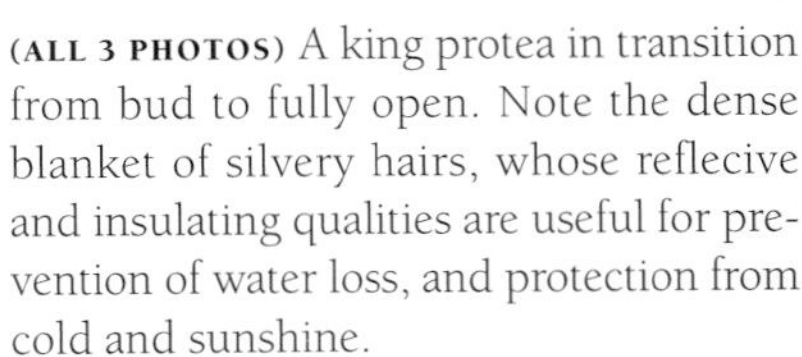

(ALL 3 PHOTOS) A king protea in transition from bud to fully open. Note the dense blanket of silvery hairs, whose reflecive and insulating qualities are useful for prevention of water loss, and protection from cold and sunshine.

(RIGHT) "Pink Mink" (*P. neriifolia*) buds off an abundance of silky flowerheads whose feathery-fringed bracts curve gently inwards at the top. **(BELOW)** "Pink Star" (*Leucospermum tottum*), is a petite, slightly fuzzy pincushion. Its flowerheads, neatly spidery and adorned with tightly rolled pink "ribbons", are especially welcome during the cooler upcountry winters.

✚ ROSE DIPLADENA

(*Mandevilla sanderi*) Apocynaceae

When I first saw this shocking pink vine, I thought the flowers were plastic... the color was so "unnatural" and there always seemed too many for the plant to be real. It is still hard for me to believe that they flowers have not been dyed. However, rose dipladena is a Brazilian, ever-blooming, woody vine which favors arbors, and seems to thrive particularly well upcountry. To attest its popularity, look for it in K-Mart!

✚ ROSE TURMERIC

(*Curcuma elata*) Zingiberaceae

Well-named, this rosy, yellow-flowered jewel is related to *'olena* (turmeric, "Green" chapter), but instead of green and white flowerheads, its upper bracts emanate a rich rose-violet color. Rose turmeric is from India. Over 50 species of *Curcuma* inhabit rainforests of Asia and northern Australia... who knows what might turn up next?

✚ SENSITIVE PLANT
(*Mimosa pudica*) Mimosaceae (Leguminosae)

Unmistakable, this low roadside weed is replete with fern-like leaves, pinkish-lavender pompons and small, rather nasty spines (watch out, bare feet!). Originally from tropical America, it binds sand well and is now naturalized throughout the tropics. Although a lowly member of Hawai'i's flora, it is known by all on account of its habit of instantly "sleeping" when touched. This is due to the sudden expulsion of watery sap from little "bladders" lying in the leaf axils. On the mainland, sensitive plant is sometimes grown in pots and gardens for its amusement value.

✚ SHOWER TREE, PINK-AND-WHITE
(*Cassia javanica*) Caesalpiniaceae (Leguminosae)

If you see a small tree smothered with spectacular clusters of pink flowers, it is most likely the pink-and-white shower. (If the pink mingles with yellows, see rainbow shower, "Yellow" or "Orange" chapters). The pink-&-white blooms early (March), followed by the rainbows (June), each lasting several months. The fern-like (compound) foliage of shower trees is typically leguminous, i.e. each leaf is subdivided into many paired leaflets.

Glorious *en masse,* the flower clusters embrace the thin branches like giant pink leis **(ABOVE LEFT)**. Individually they are also elegant: each resembles a rounded, five-pointed star of marbled pink, white, or rosy hues **(BELOW LEFT)**. The finely "painted" petal veins resemble blood capillaries, while the blossom center glows with golden stamens and a long, curved, green style. (Note that in the similar pink shower (*C. grandis*), flowers are pea-, rather than star-shaped.)

✚ SILKY JACKBEAN
(*Canavalia sericea*) Fabaceae (Leguminosae)

Appearing native, and called *pohue* in Hawaiian, this attractive coastal sea-bean is native almost everywhere in the Pacific west of Hawai'i, but like several other Pacific-wide plants (tree heliotrope, *kou, milo*), never made a successful jump to Hawai'i's isolated shores on its own. Its rosy, sweet pea-like flowers (1 1/2 inches) and long, compressed, brown seedpods (7 inches) are unmistakable. Silky jackbean dots Maui's central windward coast; try Kahului Beach Road by the harbor and Pauwela Point.

✚ SNOW BUSH
(*Brehnia distichta* cv. 'Roseo-picta')
Euphorbiaceae

Native to the South Pacific and Southeast Asia, this rounded shrub (sometimes a hedge), is best recognized by its pinkish, slightly weeping yet airy, foliage. From a distance, the oval leaves, variegated white, pink, rose, and pale green, mimic a mass of pink and white flowers or, as its common name implies, a snow-blanketed bush. The sunnier the location, the brighter the foliage. Snow bush's dainty flowers—tiny pink bells on longish stalks—are strung along one side of slender arching stems.

✚ SUMATRAN GIANT LILY

(*Crinum amabile*) Liliaceae (Amaryllidaceae)

Cultivated in Hawai'i for more than a century, and particularly tolerant to salty environments, Sumatran giant or spider lilies (see also "White" chapter) always merit a quick "fragrance stop". How artistic are the spidery, vinaceous-and-white flowers, with their erect, threadlike stamens! The spirally arranged, succulent, reddish, strap-like leaves add extra flair. This robust lily, to five feet in height and diameter, is native to Sumatra (Indonesia). Enjoy them at, Hamoa Beach, Hana **(RIGHT)** and at coastal resorts.

▼ TI

(*Cordyline fruticosa*) Liliaceae

Ti (*ki*), in its myriad varieties, is commonly grown by local residents, flourishing particularly well in moist areas. Watch for naturalized plants throughout island windward roads and forests, as well as in gardens and landscaping. Hawaiians of old used its "plastic-coated" leaves to relieve headaches. Try it. The cooling effect is soothing. Pictured is a portion of its arching flower stalk, chock full of tiny pink "lilies".

Detail of jacaranda (*Jacaranda mimosifolia*), Kula.

PURPLE FLOWERS

ALTHOUGH PURPLE FLOWERS are not especially common anywhere in the world, they sometimes occur en masse, dazzling visitors and residents alike. For example, jacarandas. Wherever they are planted for highway or city beautification, they explode into stunning springtime displays. Upcountry Maui is a prime area. This chapter also includes the striking lobelias found nowhere in the world but Hawai'i; the brilliant purple-flowered glory-bush, fast becoming a forest pest; the dazzling magenta Hong Kong orchid tree, city emblem of Hong Kong; and ravishing Dendrobium orchids **(RIGHT)**.

✚ AGERATUM

(*Ageratum conyzoides*) Asteraceae (Compositae)

A lowly member of the daisy family, ageratum, naturalized since 1871, is a widespread weed in Hawai'i's lowlands. It especially favors small watercourses, marshy areas, and roadsides along the windward coast. It is familiar to hikers. Originally from tropical America, this hairy, somewhat malodorous weed is related to the cultivated A. *houstonianum,* a delightful border plant with tightly massed flowers, now available in an array of mauve and pink colors. Recently *Ageratum* has been found to contain an oil which combats a degenerating disease of pigeon peas, an important green manure crop.

Ron Nagata/Haleakala National Park

✚ BITTER HERB OR CENTUARY

{*Centaurium erythraea (= C. umbellatum)*} Gentianaceae

One of Maui's prettiest weeds, this European import has dotted disturbed, arid areas for nearly 100 years. Occurring up to 7000 feet, it is commonest on Haleakala's south-facing slopes. Only a few inches tall, it bears small, stemless leaves and terminal clusters of purplish-pink, star-shaped flowers. The scientific name honors Centaur, a Greek half man/half horse, who evidently used *Centaurium* to heal a wound inflicted by an arrow from Hercules. Bitter herb's common name derives from ancient pharmaecopia: its tissues contain a bitter glucoside, erythro-centaurin, reputed to purify blood. It is claimed by ancient herbals to cure heartburn, although "'Tis very wholesome, but not very toothsome" (Culpepper)

✚BLUE GINGER
(*Dichorisandra thyrsiflora*) Commelinaceae

Neither blue nor a true ginger, this stunning Brazilian ornamental primarily grows in well established Maui gardens, lowland or upcountry. Blue ginger is unfortunately not readily available these days. Many people are surprised to discover that its closest relatives are spider plants and wandering jews (see "Blue" chapter) rather than gingers. Look closely at an individual flower... it's *almost* identical to a wandering jew, and certainly lacks the colored sheath-like bracts of gingers. Those bright, compact, conical flowerheads and tall ginger-like leaves fooled us all!

✚BOUGAINVILLEA
(*Bougainvillea*) Nyctaginaceae

This brilliant shrub-cum-vine beautifies the dry, relatively infertile highways in Lahaina yet also luxuriates in Hana and Pukalani. These glorious displays do not happen automatically: gardener's secrets including heavy pruning, light watering, and moderate fertilizing.

(ABOVE) A pale magenta hybrid, and **(LEFT)** detail of a rose-magenta hybrid, showing the colored bracts and tiny, white "true" flowers. See also other chapters.

Bougainvillea's common and scientific names honor Louis A. de Bougainville (1729-1811), a renowned early world navigator who discovered the original dark purple species (B. spectabilis) in Rio de Janeiro, Brazil (**RIGHT**, at Wailea Akahi).

✚ CARICATURE PLANT
(*Graptophyllum pictum*) Acanthaceae

Recently, caricature plant has become popular in resort landscaping; otherwise it is generally found in older residential areas. Although several horticultural varieties have been developed from the original New Guinean species, its distinctively variegated foliage and flowers are easy to spot. The tubular, rose-purple flowers are oddly shaped—they appear "pinched" to near-flatness, the petal edges are curled over, a pair of rolled "wings" cross over the "back", and the top lip extends forward to look like a "nose"! Every leaf looks individually painted—with *watercolors*. The commonest colors are 3 or 4 greens and yellow (note the yellow margins).

✚ CHINESE VIOLET OR ASYSTASIA
(*Asystasia gangetica*) Acanthaceae

A low, trailing ground cover with lavender-rose blossoms (also blue-purple), the Chinese violet rambles exuberantly over walls, shrubs, and grasses. The flowers are composed of five crinkly petals crowning a funnel-shaped tube about one inch long. This vine, native to Africa and Southeast Asia, is found in urban areas and moist, weedy valleys. It is also called asystasia, a word of Greek derivation meaning "loose, having no cohesion", likely referring to its peregrinatious habits. Do not confuse this species with another plant called Chinese violet or *pakalana* (*Telosma cordata*), whose yellow-green flower buds are often strung into leis.

✚ COLEUS
{*Solenostemon scutellarioides (= Coleus blumei)*} Labiatae

In general, this familiar houseplant is grown for its kaleidoscopic, variegated foliage. Gardeners often pinch back the foliage to produce a bushier, brighter effect, but its bluish-purple flowers are also attractive. Originally from Java, coleus is sometimes called "painted nettle" but it is strictly a member of the mint family, with square stems and lipped flowers. However, the similarity of its leaves to nettles is good to remember if you encounter coleus that has escaped into the wild. In shady places, roadside coleus loses its leaf color and becomes "leggy" and weak-stemmed.

✚ CROWN FLOWER
(*Calotropis gigantea*) Asclepediaceae

Not only strange and beautiful, the crown flower—whose scientific name means "giant, beautiful keel"—has myriad uses and associations with folklore, medicine, and religion. First, it symbolizes one of the arrows of Kama, Indian God of Love. So, if you are presented with a lei—fashioned from buds, the crown-like centers, or complete flowers—do be aware that such a gift is almost sacred (photo p. 69). Practically, the fibrous bark is used in its native India for tanning hides, fishing lines, and textile weaving. An enormous milkweed, crown flower's seedpods are large and overflowing with copious, kapok-like fluff—perfect for pillows. Hawai'i's famous Queen Liliuokalani loved this mauve variety (the white one, p. 100, was introduced after her death in 1917).

✚ GLOBE AMARANTH
(*Gomphrena globosa*) Amaranthaceae

Dotted in gardens you may see patches of globular, "everlasting" flowerheads that resemble giant clovers. For decades globe amaranths have been strung into leis, earning them the local name *leihua* or "ball lei" (don't confuse with *lehua*, "Red" chapter). Each "ball" is a mass of tiny flowers enclosed within dozens of overlapping colorful bracts, somewhat like an artichoke or protea. Although purple is popular, globe amaranth also comes in white, orange, or rose. Bozu is another local name.

✚ GLORYBUSH
(*Tibouchina urvilleana*) Melastomataceae

Although this is a beautiful bush, for the sake of Hawai'i's native plants struggling to survive—PLEASE DO NOT EVER PLANT IT. Even pruned branches, tossed away, will root and grow rapidly. Native to Brazil, glorybush was first brought to Hawai'i in 1910 to an estate on the Big Island. Since then it has been cultivated on all major islands and has escaped into native forests. It is a *serious invasive shrub,* wiping out increasing areas of rainforest on Kauai and the Big Island (see Environmental Alert).

✚ GOLDEN DEWDROP
(*Duranta erecta*) Verbenaceae

Aptly named, this tropical American bush/hedge/small tree is highly ornamental, bearing grape-like clusters of golden berries. Its delicate mauve flowers, each petal with a purple central band, form abundantly on thin, drooping branches, usually on one side only. They are apt to fall easily in windy areas, but the waxy, chickpea-like berries persist for months (in Colombia, it is named *garbancillo,* "little garbanzo"). *Golden dewdrop's berries are poisonous.* Although used medicinally to kill intestinal worms and in Asia for mosquito repellent, the berries contain hydrocyanic acid and a saponin fatal to children.

■ *HAHA*, TREE LOBELIAS
■ (*Clermontia, Cyanea*) Campanulaceae

Oddly curvaceous and leathery textured, the flowers of Hawai'i's native tree lobelias co-evolved with our native birds, Hawaiian honeycreepers. The ancient Hawaiians knew this, calling the lobelias *haha,* "food for the birds". As different lobelias developed, so new types of birds evolved with bird beaks to fit the new flower shapes. Thus honeycreepers were assured of a source of nectar and the lobelias were pollinated when the birds thrust their bills deep into the flower tubes.

(LEFT) This is one of my favorite lobelias. A specialty of the Big Island's wet forests is *Clermontia montis-loa,* whose 2"-long, curved flowers flare out to expose several square inches of soft, velvety magenta. It dots the understory in the Hilo, Puna, and Ka'u districts, sometimes hybridizing with *C. parviflora* to form a paler, smaller flower.

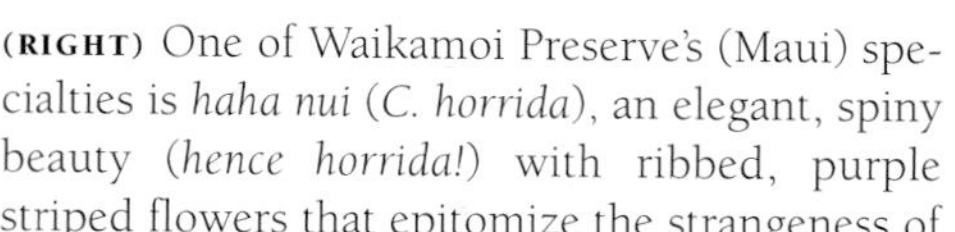

(RIGHT) One of Waikamoi Preserve's (Maui) specialties is *haha nui* (*C. horrida*), an elegant, spiny beauty (*hence horrida!*) with ribbed, purple striped flowers that epitomize the strangeness of Hawai'i's tree lobelias. *Haha nui's* entire world range encompasses only one thousand feet of elevation (5000 to 6000 feet) on Haleakala's windward slope. Its endangered status globally is G2S2.1, i.e. the majority of its plants are threatened with immediate extinction.

✚ HYDRANGEA

(*Hydrangea macrophylla*) Hydrangaceae (Saxifragaceae)

Hydrangeas, with their large, coarse foliage and complex flowerheads, are peculiar in that their flower color varies, not according to genetic makeup but by soil acidity. If the soil is acid, the flowers are bluer; if alkaline, pinker. Pink is the normal flower color, but blue/purple can be artificially produced by adding acidic substances (aluminum sulphate, peat, acid fertilizer) to the soil. Hydrangeas dot older residential areas at higher elevations, for example Kokee (Kauai), Volcano (Big Island), and Kula (Maui).

✚ IMPATIENS

(*Impatiens walleriana, I. hawkeri*) Balsaminaceae

Impatiens, with their "faces of many colors", come in the "regular" and "New Guinea" species. The latter are larger, sometimes striped, and are now highly popular with homeowners. The flowers, appearing like flat disks, actually bear long, curved, narrow spurs behind the petals. Each spur hides a relatively large amount of nectar, hidden in such a narrow tube that only butterflies or birds with very long tongues can reach it. Nectar is the "money" with which flowers pay insects for their "pollinating visit"—impatiens is evidently a generous donor.

✚ JACARANDA

(*Jacaranda mimosifolia*) Bignoniaceae

Native to Argentina and Brazil, jacarandas are medium to tall trees, thriving best at mid-elevations (2000—4000 feet). Throughout the Pacific, traditional proverbs relate to the seasonal occurrence of local fish, flowers, migrant birds, etc. If we had modern equivalents; a good candidate might be "when the jacaranda blooms, spring has come!" (substitute tax time or Easter vacation if you wish). Upcountry Maui and in the Kohala region (Big Island) are two areas in the world where jacarandas may be enjoyed in full splendor. Although the feathery foliage and floral masses resemble shower trees from a distance, they are unrelated. The flowers, for example, are tubular rather than pea- or star-shaped (see chapter heading).

✚ JAMAICA VERVAIN
(*Stachytarpheta jamaicensis*) Verbenaceae

A cheery, ubiquitous roadside weed, the Jamaica vervain is the commonest vervain in Hawai'i. A foot or two tall, it bears toothed, opposite leaves which curve quite horizontally and are not rough to the touch. The flower stalks are unbranched and composed of many tiny bracts from which only a few little purple flowers open at any one time. Jamaica vervain especially favors the windward coast. Several *Stachytarpheta* hybridize in Hawai'i, so flowers may vary in color (pinkish, bluish, violet) and leaves may vary greatly in smooth- or roughness. Related species are used medicinally to cure eye diseases, and in Uruguay, for fertility control.

✚ MADAGASCAR PERIWINKLE
(*Catharanthus roseus*) Apocynaceae

A favorite border decoration, this gay ground cover thrives in almost any soil, growing best with a little shade and plenty of water. Originally from Madagascar, its little five-petaled flowers are now widely naturalized in tropical areas. In temperate areas it is an annual. In Hawai'i, it has been naturalized since the 1870s, earning it a local name, *kihapai*. Look for it along windward roadsides. See "White" chapter.

◆ *MAUNALOA* VINE, A PACIFIC BEACH PEA
(*Canavalia cathartica*) Papilionaceae (Leguminosae)

(LEFT) Tucked away in odd coastal corners, maunaloa vines creep over sand and rocks. Native to Polynesia, it is an abundant and widespread member coastal ecosystems, but its seeds never reached Hawai'i on their own. **(BELOW)** The *maunaloa* lei style involves flattening the pea-like flowers and threading them with their keels extending outwards. However, don't give an authentic *maunaloa* lei to visiting mainland friends because a potentially injurious virus prevents exportation (Vanda orchids can be substituted). In this fancy lei, *maunaloa* intertwines with crown flower buds (this chapter), and ti leaves. Composite leis composed of several alternating spirals, like this one, are called *'oni*.

◆ MORNING GLORIES
(*Ipomoea*) Convolvulaceae

(RIGHT) Ivy-leaved morning glory (*I. cairica*), common wild in the Nahiku-Hana area, Maui. **(CENTER, RIGHT)** *Pohuehue* (*I. pes-caprae*), sprawling abundantly on beaches everywhere, is a highly effective sand-binder. Although its funnel-shaped, mauve flowers seem delicate, the entire vine, with leathery leaves and sturdy running stems, is surprisingly tough. It can withstand strong winds, scorching temperatures, and periodic saltwater immersion. Note the cleft leaves, inspiring the name *pes-caprae* ("goat's foot"). In olden times frustrated Hawaiian surfers beat *pohuehue's* vines on the sea to encourage the surf to rise. **(CENTER)** *Koali 'awa* (*I. indica*), with heart-shaped leaves, crawls over vegetation in dry areas up to about 3500 feet.

✚ ORCHIDS
Orchidaceae

(RIGHT) The purple form of the common upcountry scarlet epidendrum (*Epidendrum* x *obrienianum*), see "Orange" chapter.

(OPPOSITE, TOP RIGHT) Some of the showiest orchids are *Dendrobiums*, hundreds of which are native to southeast Asia, Australia, and the Pacific as far east as Fiji. In Hawai'i, dendrobiums are used prolifically in resorts since they grow fairly easily. The flowers, blooming in sprays, are usually purple or pink.

Orchids are not as complex as you might think. A *Dendrobium*, for instance, is characterized by two, wing-like (almost circular) inner petals and three outer, thinner sepals. The front lip is a platformed tube, and two side sepals form a back spur.

(LEFT) A *Cattleya* hybrid on Maui. Nowadays cattleya is a household word, suggestive of weddings and special occasions. Each elegant bloom has three outer sepals (narrow or fancy); two rounded, crinkly side petals, and a highly convuluted, tubular, front lip.

As you drive Hawai'i's windward roadsides, watch for the dainty Philippine ground orchids (*Spathoglottis plicata*) growing by the wayside. Unfortunately they are far less abundant now since the roadsides are sprayed with herbicides. First grown on Oahu in the 1920s, they escaped into the wild but are not pests. Note the distinctly fluted leaves (plicata means "pleated"), and spatula-like "tongues" on the flowers. This terrestrial orchid is native to southeast Asia, including the Philippines.

Perhaps the best known of all Hawai'i's orchids are the hybrid vandas, *Vanda* cv. 'Miss Joaquim'. They were grown in commercial hothouses way back in 1893. Today grown extensively for leis, they are possibly the only orchid in the world sold by the *pound* rather than by the *single flower!* Remember the days when everyone flying to Hawai'i received free vandas and champagne?

✚ ORCHID TREES

(*Bauhinia*) Caesalpiniaceae (Leguminosae)

(BELOW) Plant-watchers during the winter months are likely to be treated to a bounty of resplendent, rose-purple blossoms of the Hong Kong orchid tree (*B. blakeana*). This exquisite beauty, sporting five-inch flowers, is unmistakable. It is a hybrid discovered in Hong Kong in 1908, and now the floral emblem of that city. Like many hybrids, it lacks seeds so can only be grown by grafting.

A commonly planted orchid tree in Hawai'i (*B. purpurea*), from Asia, has fragrant, pinkish-mauve flowers which resemble those of the royal poinciana. Look for it in all major towns and the resort areas.

✚ PRIDE-OF-INDIA
(*Melia azedarach*) Meliaceae

Also called chinaberry, Persian lilac, or *'inia,* this lovely tree occurs in gardens and along upcountry roadsides. Introduced in the 1830s, it was popular for many years but is hard to obtain now. Native to Asia and Australia, it has been widely cultivated in warm regions. In early summer, branched clusters of lavender flowers decorate the attractive compound-leaved foliage. Each has five star-like petals and a purple tubular center. Collectively the blossoms emanate such a heavenly scent you can smell it from your car as you drive by. The little hard, round fruits contain a narcotic toxin which attacks the central nervous system.

◆ *POHINAHINA* (BEACH VITEX) and BLUE VITEX
✚ (*Vitex spp.*) Verbenaceae

These two coastal, mauve-flowering plants were once common denizens of Hawai'i's coastlines. **(TOP RIGHT)** *Pohinahina* or beach vitex (*V. rotundifolia*) is indigenous to Hawai'i and the western Pacific. Today it has kindly been planted as an erect sand-binder by a few beachside resorts (for example, Kaanapali beachwalk). Note the *simple, nearly stemless, egg-shaped leaves* (1-2 inches). **(BOTTOM RIGHT)** Blue vitex (*V. trifolia var. subtrisecta*), from Australasia, makes a good hedge. It is characterized by *stalked leaves divided into three pointed leaflets.* A commonly planted variety has white-margined leaflets (var.*variegata*) and miniscule flowers.

✚ ROSE MOSS OR PORTULACA
(*Portulaca grandiflora*) Portulacaceae

(BELOW) Rose moss is not a true moss nor a true rose, but a hardy member of the portulaca family native to Argentina and Brazil. Also called sun plant, it thrives in sunny garden spots, rock gardens, and slopes. It requires little water, since its fleshy leaves are well-adapted to sparse rainfall. Its showy flowers of layered petals are larger than most portulaca flowers, but true to family genetics, they only open for one day. Hawai'i harbors several species of weedy portulacas, including pigweed (*Portulaca oleracea*), which has tiny pink flowers. In Hawai'i, all small succulents (portulacas, iceplants) are called *'akulikuli.*

✚ ROSE-OF-SHARON
(*Hibiscus syriacus*) Malvaceae

Growing up in a Christian culture, the rose-of-Sharon becomes associated with the Bible. However, the original home of this mauve hibiscus bush was India and China: it was not even introduced into the Holy Land until the late 16th century! Nowadays, rose-of-Sharon grows in many tropical, subtropical, and temperate areas. It is most common at slightly higher elevations, where both single and double varieties grow on upright, sparsely leaved bushes. **(BOTTOM LEFT)** The double form is crowded with twisted petals and often the staminal tube is barely visible, so it may not be recognizable as an hibiscus.

✚ TIBOUCHINA, CANE

(*Tibouchina herbacea*) Melastomataceae

This plant has been declared an official noxious weed by the State of Hawai'i, and serious attempts are currently underway to eradicate it. Native to Brazil, Uruguay and Paraguay, it is apparantly not even cultivated in Hawai'i, and scientists do not know why it was even introduced into Hawai'i. It was first seen on the Big Island in 1979, and on Maui in 1986 behind the Boy Scout Camp, Waihee. In recent years it has proliferated cancerously on windward West Maui, and is established in Makawao Forest Reserve above Twin Falls (Hoolawa Stream). See Environmental Alert.

Randy Bartlett

Randy Bartlett

✚ WATERLILY

(*Nymphaea species*) Nymphaeaceae

Waterlilies, primarily pink or white, are cultivated throughout Hawai'i—in resorts, arboretae, condominiums, and public places. Most originated in tropical countries. Some have exquisitely pure fragrances, others unfurl their beauty only at night. Everywhere where waterlilies are native, folklore relates delightful (and often fanciful) tales relating to their origins. Unfortunately Hawai'i has no native waterlilies, but I like the Chippewa Indian belief that the first waterlily was a star that longed to be loved by children. In order to descend to earth, it allowed its reflection to dissolve in a still lake.

✚ *Heliconia stricta* cv. 'Firebird', a five foot-high heliconia grown as cut flowers and garden ornamentals. Note the yellow-margined bracts and green flowers.

RED FLOWERS

SCARLET FLOWERS BESPEAK the tropics. Red, yellow, and orange are important tropical colors because they evolved in association with a rich continental avifauna. In all warm regions of the world, birds are important pollinators. Generally, "bird flowers" have little or no scent because color is enough to attract birds.

Hawai'i glows with red flowers, native and introduced. Brilliant poinsettias brighten rainy days in winter; coral trees burst forth in spring; silverswords dress up Haleakala National Park in August, and all-year proteas and hibiscus beautify homes and gardens. Hikers and residents of forested communities can always enjoy Hawai'i's ubiquitous native *'ohi'a* trees.

(RIGHT) A dazzling bouquet of anthuriums (*Anthurium andraeanum*), red ginger (*Alpinia purpurata*), red holiday heliconias (*Heliconia angusta* cv. 'Holiday'), is offset by the silvery foliage of Indian summer banksia (*Banksia occidentalis*) and red variegated ti leaves (*Cordyline fruticosa*).

✚ ALOE
(*Aloe vera*) Aloeaceae (Liliaceae)

Although not native to Hawai'i, the cactus-like aloes have grown here for more than a century. Many *kama'ainas,* especially those who call aloes *panini 'awa'awa,* swear they have been here forever! Most aloes originated in South Africa, but this one—famous because of its medicinal and cosmetic properties—is from the Mediterranean region. Inside its spiny leaves is a greenish, toothpaste-like gel composed of specialized water-storage cells. This feature enables the plants to survive in arid environments and also withstand severe droughts. The gel promotes and accelerates healing (especially burns), also forming a protective coating which prevents further bacterial infection.

✚ ANTHURIUMS
(*Anthurium andraeanum*) Araceae

(LEFT) Anthuriums, or "love" flowers are now synonymous with Hawai'i and with Valentine's Day. Commercial production is an international, multimillion dollar industry. Since anthuriums require year-round warmth, filtered sunlight, and high humidity, most are grown in artificially shaded farms in cloudy, drizzly regions of the Big Island. Seedlings, rooted in volcanic cinders or treefern (*hapu'u*) chips, average 20,000+ plants per acre, depending on the cultivar type, planting rotation, degree of shade, and pruning schedule.

(LEFT) One of the many varieties of "tulip anthuriums", introduced relatively recently into Hawai'i. The surrounding spathe is upright and cupped, like a tulip petal, while the central spadix is straight.

✚ BANKSIAS
(*Banksia*) Proteaceae

(BELOW) "Raspberry Frost" (*B. menziesii* cv. 'Raspberry Frost'), one of the most beautiful banksias, is colored a rich burgundy, which lightens and glows as it matures. On Maui it is grown commercially and privately. Reaching 30 feet tall, each shrub grows fast, producing up to 60 blooms annually. 'Raspberry Frost' excretes copious amounts of nectar, a sugary exudate that attracts hordes of ants. They are not stinging ants, but some people might object to their presence: watch for them before you pick the flowers.

(ABOVE) Scarlet banksia (*B. coccinea*), adds a brilliant accent of coiled texture to a bouquet of rare white king proteas (*Protea cynaroides*), pincushion proteas and hanging heliconias. A native of West Australian scrublands, its looped symmetry is reminiscent of crocheting. Wild scarlet banksias are exploited heavily in Australia and conservationists are concerned for its survival.

✚ BOTTLEBRUSHES
(*Callistemon*) Myrtaceae

Bottlebrushes are hardy, drought-tolerant, fast-growing trees which are especially popular upcountry, although they do not relish hard frosts. The scientific name means "beautiful stamens", which is indeed true, since their bottlebrush-like spikes are composed almost entirely of scarlet stamens from numerous individual flowers. All are from Australia. The most common species are **(BELOW)** scarlet bottlebrush (*C. citrinus*) and weeping bottlebrush (*C. viminalis*), a taller tree with gracefully pendulous branches and delicate bronzy young foliage **(RIGHT)**.

✚ BOUGAINVILLEA
(*Bougainvillea*) Nyctaginaceae

(BELOW) Bougainvilleas with rose-red coloration (B. glabra hybrids) tend to be less thorny than the pure purple forms, which are closer to the wild Brazilian species (*B. spectabilis*, see "Purple" chapter). The reds, intermingling with other bold hues, create dazzling displays as their massive stems arch outwards over freeways and at resorts. **(RIGHT)** A double red-magenta bougainvillea.

✚ CANNA
(*Canna indica*) Cannaceae

Native to tropical America, cannas (*ali'ipoe*) are old-timers in Hawai'i. Long ago their banana-like leaves were used to wrap food, and their black seeds were strung into leis or placed in hula rattles. Some cannas are edible. Believe it or not,

they were cultivated in Hawai'i for emergency human and livestock food during World War I. Although plants started by root division bloom in a few weeks, cannas are not planted much now. Their flowers, big and bold, are easily wind-torn and wither quicker than gingers and heliconias. Canna flowers are strange: the wide "petals" are not petals at all but male reproductive parts!

✚ CARNATIONS
(*Dianthus caryophyllus*) Caryophyllaceae

Originally from southern Europe, border carnations have been cultivated for over 2000 years. They are easily propagated from cuttings. In addition to their floral beauty, extractions of their spicy-fragrant essen-

tial oils contribute to many high quality perfumes. Interestingly, the Hawaiian name is *poni-mo'i*: "coronation" (is this confusion or mistranslation?) **(ABOVE)** In Hawai'i, what graduation, wedding, or concert is complete without a double carnation lei? **(LEFT)** On Lei Day, school "princesses" representing the Big Island sometimes substitute red carnations if *'ohi'a lehua* (this chapter) is unavailable.

✚ CASTOR BEAN OR CASTOR OIL PLANT
(*Ricinus communis*) Euphorbiaceae

This plant produces the castor oil that your mother may have crammed down your throat as a child. However, the preparation secret was not to heat the shiny black seeds (oils were "cold-drawn"), since the seedcoats are so toxic that a mere 3 seeds can kill an adult! This high grade oil is today used in airplane engines, but has been used for at least 4000 years as a general lubricant. Castor bean is a weedy shrub with attractive palmate leaves (like the fingers of a hand) and large spikes of prickly round seedpods. It grows in waste places and vacant lots all over Maui. The flowers are usually not noticed, but the female ones (higher on the spike than the male) have ample red feathery stigmas.

✚ CHENILLE PLANT
(*Acalypha hispida*) Euphorbiaceae

An oddity from Southeast Asia, also called Red-hot cat's tail, this medium-sized shrub is a relative latecomer to tropical gardens (1896). Its long, velvety red flower spikes—masses of tiny flowers lacking petals—flourish in moist, hot, shady environments. Such climatic conditions are optimal in the warmest, moistest regions, for example, Hilo (Big Island), Manoa Valley, Waimea (Oahu), and Hana-Kipahulu (Maui). There is also an uncommon, white-flowered form.

✚ CORAL PLANT
(*Russelia equisetiformis*) Scrophulariaceae

Originally from Mexico, these showers of little reddish-orange bells add lacy embellishments to Hawai'i's lowland and mid-elevation gardens. Each inch-long flower, composed of a narrow tube vrowned by five short lobes, resembles a coral reef polyp. Up to four feet tall, the coral plant is preadapted to hear and dryness: it virtually lacks leaves, and photosynthesizes through its green, rush-like branches, conserving water loss. Its Hawaiian name, *lokalia*, is a translation of Rosalie, a girl's name.

✚ CORAL TREES
(*Erythrina*) Papilionaceae (Leguminosae)

Spring in the lowlands ushers in masses of red pea-like flower clusters which seem to explode from the leafless branches. Dozens of buds in each cluster mature over the following weeks, attracting nectar-feeding birds such as Japanese White-eyes and mynahs. Nectar from coral trees has high sugar and amino acid (protein) content. The Hawaiian coral tree or *wiliwili* (*E. sandwicensis*) is bears cream or apricot flowers ("Orange" chapter).

Two commonly planted coral trees are common or cockspur coral tree (*E. crista-galli*), with broad, distinctive flowers **(RIGHT)**, and tiger's claw (*E. variegata*), with more crowded flower clusters **(ABOVE RIGHT & BELOW)**. The latter not only has red flowers—its wood is red too, so much so that in Thailand a face powder is prepared from it.

✚ GINGERS
Zingiberaceae

(RIGHT) A relative newcomer to Hawai'i (1959), Indonesian wax ginger (*Tapeinochilos ananassae*) befits the age of plastics. Each flowerhead, four to eight inches long, resembles a well-matured pinecone dipped in bright glossy varnish. Indonesian (or Malaysian) ginger prefers the warm, moist, and densely shady forest, so do best in warm, moist, windward lowlands.

(RIGHT) Red ginger's (*Alpinia purpurata*) striking flowerheads, whether in flower arrangements, gardens, or landscaping, cannot be missed. Native to the western Pacific, and widely cultivated throughout the tropics, red ginger is indeed a festive plant. The true flowers—white and small—peek out from inside the flashy red bracts, as in the cultivar 'Jungle King **(CENTER, RIGHT)**. Occasionally one sees older flowerheads weighted down with baby plantlets which have germinated deep within the bracts. Since seeds rarely develop in red ginger, these aerial "kangaroo-pouches" are important means of reproduction. Dyes from the leaves produce subtle beiges and yellows.

(CENTER, LEFT) Tahitian ginger (*Alpinia purpurata* cv. 'Tahitian Ginger', is a botanical sport derived from red ginger. The flowerheads reach football size and are best seen in Hana. **(LEFT)** One of Hawai'i's oldest and still best-loved gingers is the torch (*Etlingera elatior*), pink when young and rich red when more mature, due to the emergence of many yellow-rimmed red flowers of teardrop shape (see "Pink" chapter).

✚ HELICONIAS
(*Heliconia*) Heliconiaceae

Scarlet heliconias, perhaps the most conspicuous of all tropicals, are nothing short of resplendent. Three species are common, the most brilliant of which is the hanging lobster claw (*H. rostrata,* this page). Ablaze with crimson, yellow and apple green bracts dangling boldly from a wiggly stem, it resembles a string of cooked lobster claws or parrot beaks. Even its scientific appelation, *ros-*

trata, means "beaked". Although the hanging lobster claw is a summer bloomer (it is native to the southern hemisphere subtropics in Argentina and Peru), University of Hawai'i horticulturalists are attempting to develop December blooming in efforts to create a northern hemisphere business in floral Christmas decorations. Note the yellow-green true flowers peeking out from its crab claw-shaped bracts.

(RIGHT) Red caribaea (*H. caribaea* cv. 'Purpurea'), a dazzling rainforest gem from Jamaica and Puerto Rico, is one of the largest heliconias in the world. Hawai'i's major export supply is grown in the Nahiku-Hana area but look for them also in resort landscaping at resorts statewide. These latter visitor-oriented areas are naturally semi-arid, but piped-in ground water produces a man-made lushness that favors heliconia growth.

(BELOW RIGHT) Common lobster claw (*H. bihai*), originally from Brazil and Trinidad, is so well-known in Hawai'i that it is even naturalized. Look for it along the Hana Highway–in places it creeps so close to the road, workmen slash it back with machetes.

(BELOW LEFT) A most delightful heliconia is the "Dwarf Jamaican" (*H. stricta* cv. 'Dwarf Jamaican', which grows only two feet high and bears red bracts with a pinkish sheen. It is also known locally as dwarf lobster, mini-Jamaica, and dwarf humilis. Introduced into Hawai'i in 1971, is especially amenable to small gardens and outdoor pots. Each green-margined flowerhead is only about five inches tall.

No discussion on heliconias, however brief, would be complete without mention of these two tall, elegant beauties. **(LEFT)** "Richmond Red" is a hybrid of *H. bihai* and *H. caribaea,* both pictured separately here. **(RIGHT)** "Total Eclipse" (*H. bihai* cv. 'Total Eclipse') is a rich, glowing burgundy red suffused with a lighter glow.

■ HIBISCUS, INTRODUCED AND NATIVE

✚ (*Hibiscus*) Malvaceae

(LEFT) It is natural to assume that the ubiquitous red hibiscus which adorns gardens, parking lots, and resorts (cultivars of *H. rosa-sinensis*), is Hawai'i's official state flower. However, when the proclamation was made (1923), the designated state emblem was a rare, silky petalled native, *kokio 'ula,* (*H. kokio,* **RIGHT**). Even then, many species of native hibiscus had succumbed to the ravages of cattle and forest destruction. Today *kokio 'ula* is rare, but small populations still exist in remote corners of the West Maui massif. Several varieties grow at nurseries and arboretae. In 1988, Hawai'i's state flower was changed to a native yellow species, *H. brackenridgei* ("Yellow" chapter).

(ABOVE) A decorative, variegated cultivar of the common hibiscus is "Snowflake" (*H. rosa-sinensis* cv. 'Snowflake'). **(RIGHT)** *Ko'oloa—'ula* (*Abutilon menziesii*) is another officially endangered native hibiscus whose habitat has been usurped by sugar cane and cattle pastures. Represented in the wild by pathetic, scraggly plants struggling to survive in a few dry gullies of central Maui & the Big Island, it thrives in cultivation. *Ko'oloa—'ula's* small maroon flowers bud off in small clusters amid heart-shaped leaves, whose soft downy texture indicates an adaptation to hot dry conditions.

✚ IXORA
(*Ixora coccinea, I. chinensis, I. casei*)
Rubiaceae

(LEFT) This low to medium bush, profusely covered with spherical masses of scarlet "stars", is common in low massed plantings (even by fast food takeouts). Three similar species are popular now, differing mostly in flower length. The local name for ixora is *popo-lehua*—"a ball of pompons like those of the *'ohi'a lehua* tree" (p. 89). Its true name, ixora, is derived from the Hindu god Iswara, to whom ixora flowers were offered in their Asian homelands. Because the flowers have extra long, thin floral tubes, they are perfect for butterfly pollination. Color variants include white, yellow, orange and pink. In Palau's rainforests, ixora is a beautiful flowering tree, very different from the compact bushes in Hawai'i.

✚ *LEHUA-HAOLE* OR RED POWDERPUFF
{*Calliandra haematocephala*
(= *C. inaequilatera*)} Leguminosae

(RIGHT) This shrubby Bolivian tree was named *lehua-haole* ("foreign lehua") by Hawaiians on account of its resemblance to Hawai'i's native *'ohi'a lehua* trees (p. 89). Its silky balls of pinkish-red stamens (to three inches in diameter) are offset by fern-like foliage. It grows quite slowly, especially at higher elevations. Occasionally you will see a lei made from dozens of these gloriously soft blossoms.

✚ LEUCADENDRONS, FOLIAGE PROTEAS
(*Leucadendron*) Proteaceae

(BELOW) 'Safari Sunset', a hybrid developed in New Zealand, is now a popular bouquet filler on Maui. Since the plant originated in South Africa, it blooms during our northern winter months (the austral summer) and now graces upcountry farms and home gardens. The most conspicuous parts of the leucadendron are its red "petals"—our old friends, the bracts (modified leaves). Here we unofficially treat them as "flowers", since the true flowers are huddled together inside, out of sight.

(ABOVE) 'Red Gem' is another newcomer to the protea growing areas and is grown singly or in tall hedges, especially in the lower Kula area, Maui. Its colors are less vivid than 'Safari Sunset', a smaller plant.

■ *NUKU 'I'IWI*, HAWAI'I'S "BEAKED" VINE
(*Strongylodon ruber*) Leguminosae

Only ardent hikers likely to see this special vine, but it is included here to impart to the reader yet another of Hawai'i's "hidden" treasures . . . to add further justification to protect her precious rainforests and watersheds. Of South Pacific affinity, *nuku-'i'iwi* climbs over trees and shrubs (500-2500 feet elevation) on all the major islands: one's first glimpse is a splash of vibrant color dotting the forest understory. The 3-parted leaves and long spikes of scarlet, pea-like flowers resemble those of coral trees (*Erythrina*, above). The name, *nuku-'i'iwi* means "beak of the *I'iwi*," a brilliant scarlet Hawaiian honeycreeper. Once a companion to its floral namesake, *'I'iwis* are today confined to forests above 3000 feet.

Hank Oppenheimer

✚ OCTOPUS (UMBRELLA) TREE
(*Schefflera actinophylla*) Araliaceae

Octopus trees are now grown nationwide as houseplants. In Hawai'i, they generally occur in landscaped settings, but also grow vigorously in a few wet, lowland forests (Maui, Big Island). Long floral "tentacles", complete with short-stalked "suckers", radiate from the octopus tree's crown. While the name "octopus tree" relates to the unusual arrangement of flowers, "umbrella tree", refers to the radial arrangement of the several leaflets which comprise each leaf.

■ *'OHI'A LEHUA*, A SACRED FOREST TREE
(*Metrosideros polymorpha*) Myrtaceae

No plant epitomizes the "truly Hawaiian landscape" more than the scarlet blooming *'ohi'a*. Its forests dominate Maui's—and Hawai'i's—upland. In its numerous forms, *'ohi'a* graces the arid lowlands, wet forests, and soggy bogs, ranging from plants a few inches high to majestic trees over 100 feet tall. As you admire its craggy, gnarled trunks and "shaving brush" flowers, take a long breath. To the ancient people, *'ohi'a* forests embodied the omniscient powers of creation, the gods Ku and Kane. Today, countless life forms—birds, plants, invertebrates—still depend on them for survival.

Without Hawai'i's *'ohi'a* forests, past and present, there would be no water for us to drink, no lush greenery, and little of the island's unique plants and animals. While flying over island mountains, the forests look intact but everywhere they are undermined by wild pigs and erosion.

Cameron Kepler

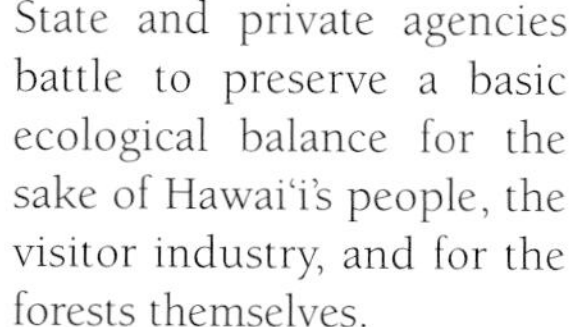

State and private agencies battle to preserve a basic ecological balance for the sake of Hawai'i's people, the visitor industry, and for the forests themselves.

(RIGHT) A beautiful modern lei of *'ohi'a lehua* in its three principal colors—salmon, red, yellow, artfully interwoven with *palapalai* ferns. (Note that *lehua* is strictly name of *'ohi'a's* flower, but in modern usage it also refers to the whole tree.) The young, pinkish-gray leafbuds are appropriately called *liko lehua. Liko* is a beautiful word with multiple meanings, including "newly opened leaf", "chief's child", and "glowing".

✚ ORCHIDS
Orchidaceae

(BELOW) *Cattleya* cv. 'Chocolate Drop' is the name of this unusually colored orchid. This particular hybrid does not fit a generalized *Cattleya* form (see "Pink" chapter), which is much frillier, especially the front "lip".

(ABOVE) Some upcountry gardens grow the small scarlet epidendrums (*Epidendrum x obrienianum*). See also other chapters.

✚ PAGODA FLOWER
(*Clerodendrum buchananii var. fallax*) Verbenaceae

This striking, large-leaved bush is quite common in older lowland residential areas. Its loosely tiered flower clusters are somewhat reminscent of Oriental temples, giving rise to the common name, pagoda flower. The scientific name, too is meaningful, "tree of fate", alluding to its varying ethnobotanical qualities. For example, in its homelands of Malaysia and Indonesia it is revered for its spiritual powers, particularly the power to summon spirits. Pagoda flower has grown in Hawai'i for long enough for it to have an Hawaiian name, *lau'awa* (leaf of the *'awa* plant), an astute observation of its resemblance to, *'awa,* a plant spiritually beneficial to Hawaiians.

✚ POINCIANA, DWARF

(*Caesalpinia pulcherrima*) Leguminosae (Caesalpiniaceae)

Long cultivated in Hawai'i, this elegant shrub was honored by the Hawaiians with the name *'ohai-ali'i*, "the royal *'ohai* (native shrubs with beautiful red, pea-like flowers)". Note the dwarf poinciana's exceptionally long stamens, an adaptation to avian pollination. Other common names are: pride-of Barbados, peacock flower, paradise flower. In addition to their beauty, *Caesalpinias* are highly utilitarian: they provide specialty items (violin bows, red dyes, tannins, laxatives) as well as timber and food.

✚ POINCIANA, ROYAL

(*Delonix regia*) Leguminosae (Caesalpiniaceae)

(BELOW & RIGHT) Another princely flower (the scientific names means "regal talons"), this is also known as flame tree or—my favorite—flamboyant. A favorite in resort and commercial landscaping, large specimen trees also abound in Manoa Valley (Oahu) and the south coast of Kauai. Mature trees in lush settings produce feathery foliage with up to 1000 leaflets in just one leaf! Originally from Madagascar, royal poinciana is now widely grown in the tropics. Deforestation in its homeland caused its near-extinction, but it was rediscovered in 1932.

✚ POINSETTIAS
(*Euphorbia*) Euphorbiaceae

(TOP RIGHT) In 1820, Joel Poinsett, the first U. S. ambassador to Mexico and an amateur botanist, became so enamored of some low Mexican shrubs with splashes of orange-red on their leaves, he shipped some cuttings back to his home in Charleston, South Carolina. Since then, both the wild type and cultivars have spread throughout the world's tropics and subtropics. Pictured is a dazzling double variety of poinsettia, which grows luxuriantly to 15 or more feet tall (*E. pulcherrima*).

(RIGHT) Wild poinsettia (*E. cyathophora*) or Mexican fire plant is not exactly the same species as was developed into commercial poisettias, but is probably similar. Watch for them in Hawai'i's lowlands in weedy wastelands or pastures, and along roadsides.

■ POKEBERRY, HAWAIIAN
(*Phytolacca sandwicensis*) Phytolaccaceae

Hawaiian pokeberry (*popolo ku mai*) is a special plant, not only because it is unique to Hawai'i, but because it is one of the relatively few island endemics of American affinity (more than 80% are derived from Southeast Asia or the Pacific). Although most of us barely give a second glance to the weedy mainland Southern pokeberry (naturalized in Hawai'i), when we encounter the Hawaiian rainforest pokeberry it is a treat.

✚ PROTEAS
(*Protea*) Proteaceae

(RIGHT) Sugar bush or honey protea (*P. repens*) formerly grew so plentifully that sailors rounding the Cape of Good Hope picked bucketfuls of their honey-drenched flowers for sweetening foods. Later it was utilized commercially for "bush syrup", South Africa's equivalent of maple syrup. Today sugar bush is grown solely for its abundant, crown-like flowers. Watch for them in upcountry gardens and flower arrangements. **(CENTER, LEFT)** A winter/spring bloomer, the duchess protea (*P. eximia*) bears large, reddish flowerheads with central mounds of regal burgundy and spoon-shaped bracts. They fade faster than other proteas, but grow fast from cuttings and can withstand cold and dryness. **(CENTER, RIGHT** 'Red Baron' (*P. obtusifolia*) resembles a smaller, more compact sugar bush.

✚ RUELLIA
{*Hemigraphis alternata (= Ruellia colorata)*}
Acanthaceae

It is surprising that this beauty from Amazonian Brazil does not yet bear a common name; ruellia is a general name for many closely related plants with red-orange bracts and/or flowers. Look for this one in botanical gardens and gardens in wet windward areas. Heavy shade does not appear to affect the brilliant scarlet bracts and fiery-orange flowers with yellow throats. Its "quilted" leaves are a glossy dark green.

✚ SHRIMP PLANT

{*Justicia brandegeeana* (= *Beloperone guttata*)}
Acanthaceae

Shrimp plant is a delightfully descriptive word, which may be simply a translation of its original Mexican appelation, *flor de camarones* ("shrimp flower"). It came to Hawai'i last century, quickly becoming an old standby in island gardens. It is not too tolerant of cold, but thrives up to about 4000 feet. Bract colors include red, orange, terra cotta, yellow, and yellow-green, but the flowers are always white with a purple herringbone design on the lower lip. Like fuchsias, shrimp plants need frequent pruning to maximize flowering and to prevent "legginess". Grow it in sunny spots as a hedge, around a mailbox, or as low massed plantings.

Ron Nagata/Haleakala National Park

■ SILVERSWORDS

(*Argyroxiphium*) Asteraceae(Compositae)

A world renowned natural masterpiece, the silversword (*A. sandwicense subsp. macrocephalum*) is unique to Maui. Another subspecies, a federally listed endangered species, occurs on the Big Island. Its center of abundance is Haleakala National Park. A natural, public display, called the "silversword loop", is located between the Visitor Center and Haleakala summit. Silverswords have evolved specialized adaptations to high elevation deserts, for example, special water-retaining leaves containing an aloe-like gel, dense layers of white hairs reflecting away harmful radiation, and a miraculous ability for their tiny seeds to germinate in barren volcanic cinders where temperatures may reach 140° F on scorching days.

Ron Nagata/Haleakala National Park

(BELOW) From July to September, many individuals—mostly over five years old—thrust up enormous flowerheads from their basal rosette of dazzling silver leaves. When fully mature, mature plants may exceed six feet high. **(RIGHT)** The hundreds of dark red daisies are embedded in sticky tissues which trap unwanted insects.

Cameron Kepler

(RIGHT) Maui boasts a celebrated endemic silversword, the Eke silversword (*A. calignis*). Unfortunately rare and vulnerable, its entire world range encompasses a spattering of summit bogs on West Maui between 4,500 and 5,500 feet. These tiny, fragile bogs are now managed for feral pigs in the West Maui Natural Area Reserve, Puu Kukui Reserve, and Eke Crater. **(BELOW)** Eke Crater and the headwalls of Honokohau Valley on an extraordinary day.

Cameron Kepler

✚ TURK'S CAP

(*Malvaviscus arboreus var. penduliformis*) Malvaceae

A common ornamental and roadside shrub throughout Hawai'i, Turk's cap's scarlet, drooping flowers resemble closed umbrellas or hibiscus that never opened fully. It has a long history of cultivation in greenhouses and as houseplants, since it flowers for months at a time and can be dwarfed, bonsai-style. Turk's cap, closely related to "regular" hibiscus, hails from tropical America, where it is also commonly planted.

Kauai's stunning endemic *koki'o ke'o ke'o* (*Hibiscus waimeae*), occasionally seen in gardens and in botanical gardens.

WHITE FLOWERS

WHITE-BLOOMING PLANTS, both native and introduced, are well represented throughout Hawai'i. They include rare native hibiscus, alpine geraniums, exquisite orchids, a "good-luck" tree, and delicate poppies which thrive in volcanic deserts.

In general, white flowers have evolved to open and/or secrete perfumes at night in order to attract pollinating moths. Some may be less visually appealing than colored ones, but many emit intoxicating perfumes which charm our olfactory senses. Who can resist smelling a white ginger, gardenia or plumeria?

(RIGHT) What classical concert is complete without a crisp white lei to top black attire? Kapalua Music Festival musicians (left to right) Richard Young, Beebe Freitas, Bill ver Muelen pose with white ginger and double carnation leis.

✚ ANGEL'S TRUMPET TREE
{*Brugmansia (= Datura) candida*}
Solanaceae

(RIGHT) The apricot variety of the angel's trumpet tree is discussed in the "Orange" chapter. The old scientific name, Datura, is derived from the Hindustani *dhatura,* the name for plants in this genus. This is unusual since most of the *Daturas* originated in South America. **(FAR RIGHT)** If you pass by the Kula Lodge, Maui at night, breathe deeply of this dramatic flower's alluring fragrance.

✚ ANTHURIUM
(*Anthurium andraeanum*) Araceae

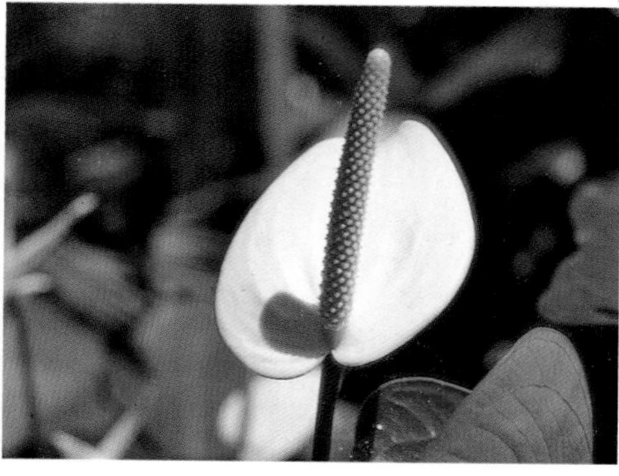

White anthuriums, less popular than the red, are sought after for weddings and for floral contrast in multicolored bouquets. All anthuriums are horticultural hybrids of the wild green species. Their kindred include jack-in-the-pulpit, calla lilies, split-leafed philodendron, and taro. See also other chapters.

✚ AUTOGRAPH TREE
(*Clusia rosea*) Guttiferae

Jacob Mau

In recent years, this superb flowering tree has become popular, especially in commercial landscaping (**RIGHT**). Look for its pink buds and large waxy-white flowers in winter and spring. In late summer and fall these have developed into round fruits which split to form beautiful star-shaped seedpods (**LEFT**). The common name derives from its thick, fleshy leaves with a thin green "skin": children like to write names on them.

✚ BALLOON PLANT
{*Asclepias (= Gomphocarpus) physocarpa*} Asclepediaceae

Originally from South Africa, this unmistakable plant was first introduced to Oahu as a fiber crop. It escaped into the wild, and although a weed, is also an attractive ornamental on account of its balloon-like fruits which resemble spiky, 2-4-inch "bladders" or terrestrial "porcupine fish". When split open, these seedpods exude masses of silky seeds. The flowers are typical milkweeds, white versions of butterfly weed (see "Orange" chapter). Look for them on Maui in the dry upper elevation pastures and along roadsides en route to Haleakala Crater.

✚ BEGONIA, WILD
(*Begonia hirtella*) Begoniaceae

This small, white-flowered begonia is familiar to all hikers in lowland Hawai'i. It is particularly prevalent along the windward flanks of Haleakala, Maui to around 3000 feet. Wild begonias favor shady, damp, disturbed locations. Look along windward highways around waterfalls, streams, wet banks, and watercourses. Native to Peru and Brazil, it was formerly cultivated, and this is undoubtedly why it escaped into the wild. Even though these days people prefer to grow larger begonias, you can find wild ones intermingled with more colorful hybrids in semi-natural gardens in outlying lowland areas.

✚ BIRD-OF-PARADISE, WHITE

(*Strelitzia nicolai*) Strelitziaceae

A close relative of the familiar orange and blue bird-of-paradise (see "Orange" chapter), the white species is much larger, often exceeding 20 feet tall). Its enormous "birds" sprout between enormous, banana-like leaves which resemble the traveler's tree (*Ravenala madagascariensis*). White bird-of-paradise plants take up plenty of room, thus display their best in large estates, resorts, botnical gardens, or in commercial landscaping.

✚ BLACKBERRY, PRICKLY FLORIDA

(*Rubus argutus*) Rosaceae

Although blackberries are fun to eat and cook into delicious jam, the ecological price paid for transitory human pleasures is great. This native of the central and east United States, with its arching, prickly stems, is a pestiferous weed in Hawai'i, especially at mid-elevations in moist areas. Particularly noxious infestations occur around Kokee, Kauai. Please do not plant them in your own yard: they will take over and then escape to become a nuisance elsewhere.

✚ BOUGAINVILLEA

(*Bougainvillea*) Nyctaginaceae

This common viney shrub is discussed in almost every chapter. White varieties, with their tiny protruding green and yellow flowers, mingle well with gaudy ones in mixed hedges. Bougainvilleas do not transplant very well but at low elevations bloom all-year. Both temperature and day length control the flowering periods: on the cooler mountainsides they become deciduous in winter. Bougainvilleas need periodic, severe pruning: don't be a picky pruner with them or you'll never get massive sprays of color.

✚ CROWN FLOWER
(*Calotropis gigantea*) Asclepediaceae

The first time you encounter this cream-flowering shrub may be via its novel flowers strung into a lei. Crown flower or giant milkweed blossoms possess a unique floral structure typical of all milkweeds, one which exhibits great specialization for pollination (see butterfly weed "Orange" chapter). Individual blossoms are composed of five, downward-pointing, twisted petals, and a crown. These floral crowns, complete with five basal curls, mimic miniature ivory royal crowns, and it is most likely for this reason that Queen Liliuokalani was particularly fond of crown flowers. Appropriately, the Hawaiian name is *pua-kalaunu* (literally "crown flower"). Leis are strung with complete flowers or crowns only. In India the buds are sacred love flowers. Kama, the god of love, shoots arrows with crown flower buds into human hearts. For purple form, see p.66.

✚ ELDER, MEXICAN
(*Sambucus mexicana*) Caprifoliaceae

For those readers familiar with mainland elderberries, elder will be familiar. For others, it is also unmistakable. Look for it along the wet highways and in gardens throughout the islands. Elder is a small tree or shrub, bearing flat-topped flower clusters (reminiscent of Queen Anne's lace) and which mature into small, edible purple-black "elderberries". The leaves are subdivided into 5-7 toothed leaflets. This species is widely planted in Central America, and botanists are not sure of its original range, but it included Texas and California. It was first collected by botanists on Oahu in 1938.

▼ ELEPHANT'S EAR
(*Alocasia macrorrhiza*) Araceae

Although elephant's ears look like giant edible taros, they should not be eaten unless you are dying from starvation and have plenty of hot water to boil the leaves and "root (underground stem) many times. Naturalized throughout Hawai'i in moist valleys and lowland forests, elephant's ear or *'ape* (pronounced "ah-pay") grows to an amazing 20 feet tall. Note the elegant, white flowers, like large calla lilies, a close relative.

✚ GARDENIA
{*Gardenia augusta (=jasminoides)*} Rubiaceae

This gloriously scented garden ornamental originated in China, but has been extensively cultivated in the western world since 1763. It is named after a Dr. Alexander Garden, who corresponded with the famous naturalist, Linnaeus. Gardenias flourish best in the warm, moist lowlands; Haleakala's south-facing slopes are really too dry. They are fussy to grow, requiring rich, acid soil and wind protection, and often develop fungus diseases. But, like roses, they are well worth the effort. Hawai'i has three native gardenias, which also smell wonderful, but unfortunately all are extremely rare.

✚ GARDENIA, CRAPE
{*Tabernaemontana divaricata (= Ervatamia coronaria)*} Apocynaceae

Variously known as crape jasmine, paper or butterfly gardenia, pinwheel flower, and Flower-of-love, this common Indian ornamental is easy to grow. It resembles a gardenia or jasmine but is unscented and not even closely allied to them ... its true kin are plumerias and allamandas! Despite this, crape gardenia's dark, glossy, paired leaves (note the deeply recessed veins of older leaves) and numerous snowy-white flowers (1-2 inches) lend themselves to fast-growing hedges, specimen plants, and shrubbery in little-watered corners of the garden. Single **(LEFT)** and double **(RIGHT)** varieties produce a range of flower forms.

✚ GAZANIAS, SILVER
(*Gazania sp.*) Compositae

Gazanias are popular ground covers. Tufted, trailing perennials, they produce masses of daisies all-year and thrive with little attention. Originally from South Africa, gazanias are also grown extensively in Southern California, from whence they came to Hawai'i. They require a frost-free climate. Various species of gazanias have been hybridized to produce many colors, but all close at night. They are useless for cut flowers and leis: enjoy them in sunny gardens, especially in drier upcountry areas. See also "Yellow" chapter.

✚ GINGERS
Costaceae, Zingiberaceae

(RIGHT) Posing as a spiralled "bamboo", crepe ginger, spiral flag or Malay ginger (*Costus speciosus*) grows wild along windward highways and occupies shady corners of public and private gardens. Its large, frilly flower bruises easily, therefore this ginger is not grown commercially. The single, fluted "petal" is not a petal at all but part of the male reproductive system. Crepe ginger's rhizomes harbor potent steroids used for birth control pills in India: perhaps, like verbenas, it may be an important medicinal plant of the future.

(CENTER) The intoxicatingly sweet aroma and delicate orchid-like configuration of white ginger (*Hedychium coronarium*) make this one of Hawai'i's favorite flowers. You are special indeed if someone gently places a white ginger lei around your neck: they are time-consuming to make, expensive, and carry an aura of deep respect and personal *aloha*. Originally from tropical Asia, white ginger's blossoms evolved a sweet, gingery nectar intended for moths and butterflies, but delicious to humans too. Look for them wild along windward highways or in well-watered gardens. The mellifluous Hawaiian name is *'awapuhi ke'oke'o,* literally "white ginger".

✚ GUAVA
(*Psidium guajava*) Myrtaceae

Introduced into Hawai'i in 1791, guavas (*kuawa*) have spread widely, creating enormous acreages of alien forests. A pestiferous shrub, its seeds are spread by cattle, pigs, horses, and frugivorous birds. The decaying fruit provides ideal breeding grounds for fruit flies. Guava dominates most of Hawai'i's moist valleys and lowland areas. Fortunately the egg-sized yellow fruits **(RIGHT)** are tasty and rich in vitamin C. They also a source of vitamin A, calcium, and iron. Guava's flowers **(LEFT)** have rounded petals and numerous fluffy stamens. California fruitgrowers are *petrified* of fruit flies, *so don't even think of smuggling guavas past agricultural inspectors*

■ *HAHA*, TREE LOBELIAS
(*Cyanea*) Campanulaceae

From chance dispersal of tiny seeds from South America, lowly blue-flowered herbs evolved riotously in Hawai'i, evolving into a variety of rosetted life-forms up to 40 feet high. They eventually occupied every habitat: seashore, dry leeward forests, wet windward forests, alpine bogs, grasslands, and precipitous cliffs. It is lamentable that today most of our fabulous Hawaiian lobelias face bleak futures. Many biologists come to Hawai'i with no other purpose than to see them. **(TOP LEFT)** Only found in Haleakala's rainforests between 3100 and 4800 feet, *Cyanea aculeatiflora* is one of Waikamoi Preserve's specialties.

■ *HA'IWALE*, NATIVE "AFRICAN VIOLETS"
(*Cyrtandra*) Gesneriaceae

Ha'iwale are rather delicate small shrubs that may be seen by hikers in Maui's wet mountains and valleys, for example, Waikamoi Preserve. Of New Guinean and South Pacific affinity, the group is exceedingly complex in Hawai'i: over 300 endemic species and numerous natural hybrids are currently described. Pictured are **(MIDDLE LEFT)**the streamside *C. grayi,* fairly common on Molokai and Maui and **(BOTTOM LEFT)**the velvety *C. platyphylla* (with its own name, *'ilihia*) from mid-elevation forests. The Waikamoi area is the best location in the world to see *'ilihia,* since its Big Island population is very rare.

✚ HEAVENLY BAMBOO OR NANDINA
(*Nandina domestica*) Berberidaceae

Locally known as the Japanese goodluck plant, this small shrub has been cultivated in Japan for centuries; plant it to the left of your front entrance for good luck. (For a double dose of lucky charms, we have ti beside our right entryway too.) Its terminal clusters of long-lasting white flowers are followed by bright red berries which provide Christmas garden color.

■ HIBISCUS, WHITE
✚ (*Hibiscus*) Malvaceae

Whenever I encounter white hibiscus blossoms, I always smell them. Fragrance is rare among hibiscus but Hawai'i's two native ones are exceptions. Their large size and exquisite perfume have long endeared them to horticulturalists. Many beautiful hibiscus in the islands are sweet-smelling hybrids with part-Hawaiian "blood". **(RIGHT)** *Koki'o ke'oke'o,* Kauai's's white hibiscus (*H. waimeae,* see also chapter heading) is uncommon and localized in the wild but blooms profusely in private gardens. **(BELOW)** A white form of the common hibiscus (*H. rosa-sinensis* 'Swan Lake').

✚ JASMINES & JESSAMINES
(*Cestrum, Jasmimum*) Oleaceae, Solanaceae

The name "jasmine", derived from the Persian *yasmin,* means "white flower". For centuries they have been revered for their delectable aromas, although some species are scentless. Approximately a dozen ornamental jasmines grow in Hawai'i.

(ABOVE) Night-blooming jessamine (*C. nocturnum*) is popular upcountry. Its fragrance is strong and sweet, heavily pervading the night air. Don't plant it unless you are aware of its smell, since its numerous seeds are very viable. The Hawaiians call it *'ala-aumoe,* "fragrant at night". (See also under "pikake", this chapter.)

(LEFT) Spanish jasmine's (*J. grandiflorum*) long viney stems bud off divided leaves and fragrant, purple-tinted flowers. In France it is grown for the perfume industry. Humans, incidentally, are not the only animals attracted to jasmine scent: a bee, with an olfactory acuity similar to our own, can detect jasmine scent at dilutions of 1:20,000!

(LEFT) Star jasmine (*J. multiflorum*) produces a year-round abundance of snowy, star-like blossoms with quite narrow petals (4 to 9). Originally from India, it is a bushy climber with slightly fuzzy leaves. After the petals fall, the calyx is a good identifying character: each lobe is divided, fan-like, into many thin green branchlets. Flowers are not always fragrant.

(RIGHT) Angelwing jasmine {*J. laurifolium (=J. nitidum)*},native to the Admiralty Islands, southwest Pacific, is identified by its smooth stems, glossy ovate leaves, dark pink buds, and star-like flowers with nine narrow petals.

▼ *KAMANI,* A PACIFIC HARDWOOD (*Calophyllum inophyllum*) Guttiferae

This handsome tall tree, typical of western Pacific coastlines, is revered in many island cultures. Like tree heliotrope, *hau,* and *milo, kamani* could be indigenous to Hawai'i, but since it never reached these islands on its own, it was brought here by early Polynesian settlers. It was used medicinally by Hawaiians, and the richly colored, fine-grained wood was also favored for calabashes. In the last few years it has been planted in a few shopping malls and resorts. Even when not blooming, the very close-veined, leathery leaves and smooth green, walnut-sized fruits are characteristic.

✚ *KAMANI,* FALSE OR WEST INDIAN ALMOND TREE (*Terminalia catappa*) Combretaceae

Abundant throughout Hawai'i's shorelines (especially to windward), false *kamani* is known to almost everyone, but don't confuse it with the unrelated true *kamani* **(ABOVE)**. Its large, egg-shaped leaves (often orange or red) and green, elliptical seedpods (1 to 1 1/2 inches long) are usually noticed before its six-inch spikes of tiny white flowers **(RIGHT)**.

(LEFT) The north coasts of Kauai (Hanalei, Lumahai), Oahu (Waimea and eastward), and Maui (Hana Highway) are excellent spots for false *kamani.* A spreading tree native to Southeast Asia, it is also called tropical almond on account of its edible "almonds", ensconced deep inside the seedpods.

■ *KANAWAO*, A NATIVE HYDRANGEA
(*Broussaisia arguta*) Hydrangaceae

Ron Nagata/Haleakala National Park

We generally associate hydrangeas with temperate, cool climates, and this applies to most family members. Thus Hawai'i's endemic *kanawao,* distantly related to a Malaysian shrub, is most unusual, especially since it is not restricted to high elevations. It is familiar to hikers in Hawai'i's rainforests (1000—6500 feet); Alternate Hawaiian names are *kanawao ke'o ke'o* ("white *kanawao*") and *kanawao 'ula'ula* ("red *kanawao*"). They reflect astute ancient observations that the shrub's bears two types of flowers: whitish-green (male, **RIGHT**) and pinkish-green (female). Fruits are small red berries.

■ *KAPANA*, A RAINFOREST MINT
(*Phyllostegia bracteata*) Labiatae

Eons ago, seeds from ground-dwelling mints somehow traversed 6000 miles—perhaps in a high jet stream—from Malaysia to the Hawaiian Islands. These oceanic waifs evolved into an endemic group of over 100 species of rainforest herbs, all of which have the square stems and "lipped" flowers of typical mints but no minty fragrance. Unfortunately many are now extinct or nearly so. This forest floor *kapana,* an endangered species, blooms amid a profusion of native ferns, orchids and other understory plants in forests above Maalaea, Maui.

■ *KAWA'U*, A NON-SPINY HOLLY
(Ilex anomala) Aquifoliaceae

A member of the holly family, *kawa'u* is a common rainforest tree on all the major islands except Lanai. Its dark, leathery leaves are unmistakable, since they have distinct veins meshing like a fine fishing net. Bunches of white, waxy flowers with yellow centers mature into black, pea-sized berries. In old Hawai'i, *kawa'u's* grayish-yellow wood was prized for carving into anvils for beating tapa—evidently it made a booming sound.

✚ *KOA-HAOLE*, AN ABUNDANT FOREIGN TREE
(*Leucaena leucocephala*) Leguminosae

Native to tropical America and first collected in Hawai'i in 1837, this shrubby tree is useful as protein-rich cattle fodder, for soil improvement (all legumes add nitrogen to the soil), and for seed leis. In 150 years *Koa-haole* Shrubland has proliferated greatly in Hawai'i's wastelands (formerly dry forests), and is common throughout the islands. *Koa-haole* (pron. "koah-howlee") is superficially similar to *kiawe,* another legume prevalent in dry areas. However, *koa haole* lacks spines, and bears white pompon flowers instead of the long greenish-yellow spikes on the spiny *kiawe* (see "Yellow" chapter).

✚ KOSTER'S CURSE
(*Clidemia hirta var. hirta*) Melastomataceae

This hairy, small-flowered shrub has for many years been classified as an official noxious plant in the State of Hawai'i (see Environmental Alert, front pages). It is a serious pest on Oahu, where it covers more than 250,000 acres. In 1961 its distribution was restricted to Oahu, now it is on all islands. So far it is not on West Maui. Biological controls are being sought; on Oahu its leaves are generally very ratty from the gnawings of control insects. Koster's curse produces abundant seeds, disseminated by birds and wind and, when fallen, by hikers and feral pigs. Please wash boots after hiking.

▼ *KUKUI* OR CANDLENUT TREE
(*Aleurites moluccana*) Euphorbiaceae

Originating and Asia and transported by Polynesian seafarers throughout the Pacific, the *kukui* (pron. "koo-koo-ee") is today one of Hawai'i's most characteristic and loved trees. Its major traditional use was for ingenious lighting devices: its nuts contain a high-grade oil like linseed oil (hence "candlenut tree"). Tannin from the bark once tanned hides and fishing nets, and the sap was a "cortisone" ointment for rashes. *Kukui* Forest is common throughout the islands, especially in windward forests. *Kukui's* small, greenish flowers were woven into leis and also provided a mouthwash. If floating in streams, they signified bad weather coming from the mountains. Note the maple-like leaves. *Kukui* is the official flower of Molokai.

✚ LOQUAT
(*Eriobotrya japonica*) Rosaceae

There are few members of the rose family in Hawai'i, and almost all are introduced like the loquat, whose seeds came in the pockets of Japanese immigrants. Today you are most likely to see loquats in areas with Japanese homes, especially the cooler areas upcountry. Occasionally they grow semi-wild along roadsides, at times in fairly thick hedges (Kula, Maui). Originally from China, the loquat is recognized by its dense rusty hairs blanketing flower buds and leaf undersides. The five-petalled flowers are rose-like if you look closely. The fruits—if they are not pre-pecked by white-eyes—are delicious raw or in jam.

■ *MA'OHI-OHI*, A VINEY FOREST MINT
(*Stenogyne kamehamehae*) Labiatae

This beautiful native vine is surprisingly common in forests of Maui and Molokai (3,500—4,500 feet), attracting native birds such as *'Apapane,* Crested Honeycreeper, and *'I'iwi.* The curvaceous tubular flowers, about 1.5 inches long, may also be pink or dark red (Molokai only). *Ma'ohi-ohi* (20 species) are endemic to the Hawaiian Islands; Maui has the most species. The Nature Conservancy and National Park Service are managing their habitats, and in some areas (high elevations, Big Island and Maui), they have experienced a remarkable comeback due to feral mammal control. Members of the mint family are easily recognized by their square stems and "lipped" flowers.

✚ *MARUNGGAY* OR HORSERADISH TREE
(*Moringa oleifera*) Moringaceae

This leaves, flowers, and seedpods of *marunggay* tree look so much like a legume, anyone but the most astute botanists would be surprised to find that it is not! It belongs in a very small, little-known tropical family. On Maui, *marunggay* is most closely associated with Filipinos, hence I use their name here. Originally native to India, it is raised commercially in Asia. Although called "horseradish tree", it is not the source of that condiment with we are most familiar. Its roots do have a horseradish-like piquancy and are used as condiments in Asia. *Marunggay* is a lanky, upright tree dotted around residential areas favored by Filipinos. Its foot-long, bean-like pods make a good vegetable, and when mature, produce a valuable oil which never goes rancid—"watchmaker's oil."

✚ MONSTERA

(*Monstera deliciosa*) Araceae

Monstera (or swiss cheese plant) is grown as an ornamental and houseplant throughout tropical regions. A majestic woody vine that begins its life as a seedling in dense forest shade, it climbs and clings to trunks until it is 100 or more feet above the ground. Of Central American origin, it is today rare in the wild because of widespread deforestation. Its large, arum lily-like flowers, resembling those of elephant's ears **(RIGHT)**, mature into phallus-like fruits with pyramidal protuberances—like a pineapple but cylindrical—which have a delicious "tropical fruit" flavor when they are ripe and mushy (hence the name *deliciosa*). Monstera's leaf holes are caused by genetically programmed dying of the leaf surface: certain patches of cells dry and split as the rest continues to grow **(RIGHT CENTER)**.

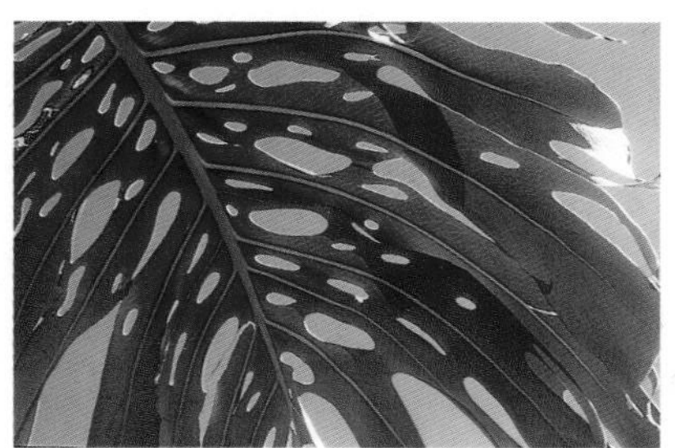

■ *EH NAIO* OR FALSE SANDALWOOD

(*Myoporum sandwicense*) Myoporaceae

Naio Shrubland once covered sizable areas of all Hawai'i's main islands: only scraps remain. Once the 'Ewa coastal plain (Oahu) was dominated by shrubby *naio*. Today, the best areas are Mauna Kea Forest Reserve & Ka'ohe Game Management Area (Big Island, 5000-7000') and the native coastal trail at Wailea Point (South Maui). *Naio* is also called bastard (false) sandalwood, because its heartwood smells similar to sandalwood, but its quality is decidely inferior. After Hawai'i had been stripped of its true sandalwood, many loads of *naio* logs were shipped to China, only to be rejected. In old Hawai'i, *naio* timber was important in house construction, while the offcuts were carved into netting needles or shuttles. The now-extinct Kona Grosbeak (*Chloridops kona*) fed almost entirely on *naio* fruit.

NATAL PLUM

✚ (*Carissa macrocarpa*) Apocynaceae

In Hawai'i, natal plum seems to be a favorite hedge of landscape architects who design shopping center parking lots. Its one-inch, star-like flowers are delightfully fragrant. Their stark whiteness contrasts with the dark green foliage, which is so tightly interwoven with thorny twigs and branches it is impenetrable. Natal plum's cherry-like fruits are edible. Native to the warm areas of Natal, a province of South Africa (where its colorful native name is *amatungulu*), it is a spiny, drought resistant shrub which tolerates neglect and poor soils. No wonder it is a parking lot specialty! Natal plum is hard to miss.

■ *NAUPAKA*, NATIVE SUCCULENTS
◆ (*Scaevola*) Goodeniaceae

True to its Hawaiian name, *naupaka-kahakai* ("*naupaka*-by-the-sea"), beach *naupaka* or scaevola (*S. taccada* or *S. sericea*) is a true denizen of Pacific coasts. Formerly abundant along Hawai'i's coasts, it is less common now **(BELOW)**. Fortunately in recent years naupaka's high salt- and wind tolerance, and ability to stabilize sand, have been recognized. **(RIGHT)** It is now popularly planted along beaches. You cannot miss it at Kilauea Point (Kauai). Beach *naupaka's* "half-flowers", famous in Hawaiian legend, are not "flowers torn in half" but complete flowers divided by a deep slit.

Listed as endangered, false jade plant (*S. coriacea*) created quite a stir on Maui in the late 1980s when efforts to protect its last sizable population were overridden by construction interests. **(BELOW)** These healthy vines were photographed at Wailea Point, South Maui, the best spot in the world to view them. **(RIGHT)** Only a handful of scraggly creepers remain on Maui County's coasts and offshore islands.

Over millenia, seeds and plants of the succulent beach *naupaka* gradually progressed away from Hawai'i's coasts into the rainforests, evolving into 7 mountain-adapted species. These were named *naupaka kuahiwi* ("*naupaka*-of-the-uplands") by the Hawaiians, who recognized their true botanical affinity to "*naupaka*-by-the-sea" (beach *naupaka*). Oahu's mountain *naupaka* (*S. glabra*) is a small tree with white flowers and non-succulent leaves. Once *naupaka* changed from a beach to montane rainforest habitat, water-storing leaves were no longer necessary: they subsequently evolved towards smaller size and pointed tips, enabling them to shed excess water. On other Hawaiian islands the mountain *naupaka* bear pink, yellow, purplish-brown, purple, or white flowers.

▼ *NONI* OR INDIAN MULBERRY
(*Morinda citrifolia*) Rubiaceae

A medicinal staple throughout the Pacific, this glossy-leaved, white-flowered, small tree may be seen in *kama'aina* homes, in the "cultural plants" sections of arboretae, and wild in lowland forests. Medical researchers have found that *noni's* pear-like fruits contain an active healing ingredient, "morindin" ... this is no surprise to Pacific islanders, many of whom still swear by it for a variety of ailments. For example, on Chuuk (Truk) Atoll, Micronesia, *noni* fruit are eaten to lower cholesterol levels. There is one catch though—the foetid, ripe fruit smells and tastes nauseating. Noni is common in lowland forests, especially close to ancient Hawaiian sites. In old Hawai'i, *noni* roots furnished red and yellow dyes, and mashed fruits were used to kill *uku* (head lice).

✚ OLEANDER
(*Nerium oleander*) Apocynaceae

White oleanders, less common than the various pink forms, are nonetheless attractive. Their simple, five-petalled blossoms emanate a sweet fragrance that, like plumeria, is not easily forgotten. Introduced into Hawai'i last century, oleanders have been cultivated in their native southern Eurasia and other frost-free areas since 1596. Multibranched, evergreen shrubs (to 20 feet tall), today they enliven gardens, roadsides, parking lots, resorts, office complexes, and airports. With full sun and adequate water, oleander blooms profusely all-year, more prolifically than hibiscus. Remember that the milky sap is poisonous (see "Pink" chapter). The Hawaiian version of oleander is *'oleana*.

✚ ORCHIDS, INTRODUCED AND NATIVE
■ Orchidaceae

Phalaenopsis, or "moth orchids", are usually white or pink. All are epiphytes (perching plants), originally from tropical rainforests of Australasia. Since they grow fairly easily, *Phalaenopsis* orchids are used extensively for indoor ornament. **(RIGHT)** A gently arching spray of *Phalaenopsis* cv. 'White with Red Lips'. **(BELOW, LEFT)** A plethora of superb Angraecum hybrids have been developed from about 200 wild species native to Africa and islands in the Indian Ocean. They are usually white with slender, trailing spurs. Pictured is *Angraecum giryamae x A. comorense.* **(BELOW, RIGHT)** Despite the fact that 15,000-20,000 orchid species exist, mostly from the continental tropics, Hawai'i can only claim a meager three species which—honestly—really only excite botanists and biogeographers. All are threatened with extinction due to habitat degradation. *Liparus hawaiensis,* shown, is rare in wet forests on all the main islands except Lanai. It resembles several species in Southeast Asia, its ancestral home.

John Carothers

✚ *PAMAKANI,* MAUI, A NOXIOUS WEED

(*Ageratina adenophora*) Asteraceae (Compositae)

Also known as *pamakani-haole,* this is a tenacious weed of dry and wet regions (2000—6000 feet). Hikers know well its unpleasant odor and tiny filamentous flowers, which resemble *Ageratum* (see "Pink" chapter). Native to Mexico, it was introduced to Hawai'i at Ulupalakua (Maui) in 1885 (hence the name's association with Maui), after which it spread to all major islands except Kauai. In open areas and shrublands, both wet and dry, it developed into such a pest, cattle ranchers were distraught. In 1944 biological control measures were begun in the hopes of curtailing its rampant spread: a parasitic fly was brought in, whose maggots cause the numerous galls which today are characteristic of *pamakani.*

✚ PAPAYA

(*Carica papaya*) Caricaceae

Papayas arrived in Hawai'i soon after Western man's arrival. Today they are cultivated extensively and locally naturalized, often along roadsides in wet areas. Originally, papaya plants came in three sexes: male, female, and hermaphrodite (male and female flowers on the same plant). At best, only 66 percent of planted seedlings bore fruit. The "useless" male plants were always so disappointing! Fortunately Hawai'i's horticulturalists came to the rescue, developing delicious, stringless 'Solo' papayas: every seedling bears fruit. Non-cultivated papayas usually bear flower clusters with short stalks (female, **BOTTOM CENTER**) or long (male,**TOP CENTER**). Smell the male flowers.

✚ PAPERBARK

(*Melaleuca quiquenervia*) Myrtaceae

Melaleuca is a forestry tree, naturalized in disturbed, wet forests on all the main islands. With its long-elliptical leaves and cylindrical spikes of creamy, stamen-filled flowers, it looks like a hybrid between bottlebrush and eucalyptus. Native to Australia, New Guinea, and New Caledonia, it first arrived in Hawai'i in 1920. Since then, nearly two million trees have been planted in forestry plots on all the major islands. It is an aggressive colonizer, thus its seeds have spread into lowland forests, usurping native species and becoming a pest. Melaleuca is easily identified by its pale, spongy bark which peels and shreds off the trunk like some species of eucalypts.

✚ PASSIONFRUIT OR *LILIKOI*
(*Passiflora edulis*) Passifloraceae

Originally from Brazil, this attractive passionfruit vine with curious purple-ringed white flowers and delicious yellow or purple fruit, came to Hawai'i in 1881 via Australia. Seeds were first planted in the District of Lilikoi, an area of East Maui encompassing Kokomo and Haiku. Today Maui residents regularly call the fruit *lilikoi* rather than "passionfruit." The English name refers—maybe to the disappointment of some readers—not to magical love potions but to diverse aspects of Christ's crucifixion: Holy Trinity, five wounds, crown of thorns, binding cords, Star of the East, spearhead, 30 pieces of silver, nails, etc.

✚ PEPPER, BIRD OR RED-HOT PEPPER
(*Capsicum annuum*) Solanaceae

Bird pepper is a well-known shrub (to 8 feet tall) grown in gardens or patio pots. The little white, tomato-like flowers grow in ones or twos, maturing into red hot peppers (less than one inch long) with a real *picante* bite to them. This wee pepper, *nioi,* was an early introduction to Hawai'i (1815). Make sure you don't pick some and put them in your shirt pocket, then hug a friend .. you'll feel the biting heat on your chest. Many other shapes and sizes of peppers are grown in Hawai'i, all varieties of the same species.

✚ PERIWINKLE, MADAGASCAR
(*Catharanthus roseus*) Apocynaceae

Native to Central America and introduced very early to Madagascar by sailing ships, this well-known ground cover has made news headlines: some of its 60+ alkaloids were found to be particularly effective in treating diabetes and leukemia. Deforestation is severe throughout the world these days, resulting in the rarity or extinction of many unique plants and animals. Already, scientists have lost chances to test other plants for their possible beneficial effects to mankind. Another common cultivar is mauve-pink (see "Purple" chapter).

✚ *PIKAKE* OR ARABIAN JASMINE
(*Jasmimum sambac*) Oleaceae

Pikake is such a well-known garden shrub/hedge in Hawai'i, it is treated separately from other jasmines (pages 104-105). There are two kinds of flowers, single and double, both deliciously fragrant. The single *pikake* **(RIGHT)** is crisply white, with a narrow tube bursting into about 9 petals at the top.

The double is quite different, having dense, rose-like blossoms packed with petals and unfolding from purplish buds. *Pikake* ("peacock") was named by Princess Kaiulani because she loved both peacocks and this jasmine! This is the species used for Chinese jasmine tea: 40 lbs of flowers are equired to scent 100 lbs of tea.

✚ *PIKAKE*, FALSE OR GLORY TREE
{*Clerodendrum chinense (=fragrans)*} Verbenaceae

The flowers of false *pikake* resemble double pikake, even to the pinkish-purple buds and calyx, but the leaves are very different. *Pikake* leaves are around one inch long, roundish, and downy (slightly fuzzy), whereas in the false pikake they are 5 or 6 inches long, broadly oval, and pointed. Look for false *pikake* shrubs in pastures and gardens near Hana (Maui), and along the north coast of Kauai.

✚ PLUMERIAS
(*Plumeria rubra, P. obtusa*) Apocynaceae

Perhaps Hawai'i's best loved flower, the white plumeria's simple, star-like blossom (*pua melia*) overflows with an irresistably sweet fragrance. Introduced last century, they were originally confined to graveyards and shunned as garland flowers because of this macabre association. To make a single lei (*kui pololei*), pick 50-70 flowers and thread a steel sewing (or lei) needle with crochet, carpet, or #10 cotton thread. Insert the needle into the plumeria's open end, and pass it completely through the tube. Repeat until all flowers are used. Tie the ends together when the lei is about 40 inches long. If you give it to someone, don't forget a kiss on the cheek. See "Pink", "Yellow" chapters.

The local name, Singapore plumeria (*P. obtusa*), is interesting and a half-truth! Long ago a plant of *P. obtusa* was taken from the West Indies to Singapore. In 1931 a cutting was brought from Singapore to Hawai'i, after which the plant became known everywhere as "Singapore plumeria". It is easily distinguished from "regular" plumerias (primarily *P. rubra*) by its glossy, narrowly oblong leaves with rounded tips and habit of blooming all-year. The flowers are always large and white (with yellow centers), and the leaves are evergreen (i.e. not deciduous).

✚ PROTEAS
(*Protea*) Proteaceae

(RIGHT) Clothed in white fluffiness, a budding queen protea (*Protea magnifica*) will eventually unfold into a rosy or chartreuse globe of fur-tipped bracts. In its native South Africa the "queen" inhabits wind-whipped, craggy mountains where icy winds and snowy blizzards are the norm. Its original names, "giant woolly-beard" and "woolly-headed protea" better describe its "alpine clothing", a necessary adaptation to survival. **(BELOW)** In the "Pink" chapter we discussed the usual form of king proteas (*P. cynaroides*). Occasionally they occur in a glorious snow-and-ivory form.

✚ PSEUDERANTHEMUM, YELLOW-VEIN
(*Pseuderanthemum carruthersii var. carruthersii*)
Acanthaceae

This variegated-leaved hedge has become popular in recent years, especially in commercial landscaping and around resorts. Its yellow leaves are reminiscent of crotons, but the distinctive flowers make this one easy to identify. Look for upright sprays of four-petalled, white flowers with purple, orchid-like central patterns. Yellow-vein pseuderanthemum (sometimes called eldorado) prefers full sun for compact growth and plentiful blossoms; in the shade it becomes lanky and less colorful. It is one of the relatively few ornamentals native to South Pacific islands but is not indigenous to Hawai'i. A close relative has purple-and-yellow variegated leaves and magenta flowers (*P. carruthersii var. atropurpureum*).

■ *PUA KALA*, AN HAWAIIAN POPPY
(*Argemone glauca*) Papaveraceae

(below) Poppy flowers appear delicate, but they surprise us by popping up in arid Californian plateaus, wind-swept Alaskan tundra, and in Hawai'i's lava deserts. Hawai'i's native prickly poppies, *pua kala* (literally "spiny flower") sparsely dot leeward slopes of all the major islands except Kauai. **(LEFT)** Deforestation and cattle ranching have lamentably reduced it to an endangered species. An invaluable addition to ancient Hawaiian pharmacology, its crushed stem, sap and seeds were mixed to relieve toothaches. Since opium is a renowned anaesthetic, this property is not too surprising.

✚ *PUA-KENIKENI*, A FRAGRANT ASIAN TREE
(*Fagraea berteroana*) Loganiaceae

Another South Pacific native, *pua-kenikeni* is one of my favorite trees. The landscapers haven't found it yet, so look for it in established residential areas: Honolulu, Hilo, Kokee, Kahului. It is sacred in Tahiti. The creamy-white blossoms, appearing in clusters at the ends of branches, fade to a rich yellow, all the while emanating an intoxicating scent. On other Polynesian islands, the blossoms sometimes scent coconut oil. In Hawai'i, *pua-kenikeni* is occasionally incorporated into leis. The local name means "ten-cent flower"; the flowers were evidently once expensive.

Ron Nagata/Haleakala National Park

■ *PU'E*, A RARE BOG LOBELIA
(*Lobelia gloria-montis*) Campanulaceae

Around 1990, the addition of The Nature Conservancy's Kapunakea Preserve contributed to an uninterrupted parcel of 13,000 acres of protected land in West Maui's mountainous massif. Among the special plants undergoing protection is *pu'e*, a spectacular rosetted native lobelia whose bogs require frequent surveillance on account of the destructive activities of feral pigs. The largest populations of *pu'e* occur on West Maui, confined to a relatively small acreage of cloud shrouded, alpine bogs. L. kauaensis occupies similar boggy habitats on Kauai.

■ *PUKIAWE*, A COLORFUL ALPINE BUSH
(*Styphelia tameiameiae*) Epacridaceae

Named after Hawai'i's king, Kamehameha the Great, *pukiawe* is the most characteristic plant of Hawai'i's high elevation alpine expanses: Haleakala National Park, Maui and cinder deserts of the Big Island. It also grows in forests, where it is less prickly and taller. *Pukiawe* is best known for its plethora of pink, red, or white berries (**BELOW & RIGHT**). These are inedible to humans but staple to the *nene's* (Hawaiian Goose).

In spring the low bushes are covered with tiny bell-like flowers. Although *pukiawe* is considered "Hawaiian", it is also native to the Marquesas Islands, French Polynesia. Its little-known family occurs primarily in New Zealand and Australia.

✚ PURPLE PLAGUE OR VELVET LEAF
(*Miconia calvescens*) Melastomataceae

This is Hawai'i's most dreaded introduced noxious plant, a rainforest tree growing to 60 feet tall and which thrives in both sun and shade. It is capable of completely exterminating all of Maui's ecosystems, including birds, insects, and plants. This has severe consequences for our precious watersheds as well. Tahiti is fast losing all its native plants and animals (see Environmental Alert). Please alert authorities to any specific locations, since active eradication is ongoing. Call 548-5250 (Oahu), 572-1983, 871-5656 (Maui), 933-4447 (Big Island), 567-6150 (Molokai), 241-3413 (Kauai). Purple plague's white flowers are insignificant, but the large, velvety leaves, shiny green above and bright purple beneath (to 2 feet long), with 3 bold leaf veins, are unmistakable. The most likely areas are the Hana Highway to Kipahulu (Maui), and north of Hilo (Big Island).

■ SILVER GERANIUM, A NATIVE ALPINE SHRUB

(*Geranium cuneatum*) Geraniaceae

(BELOW) Visitors to Haleakala National Park, Maui cannot fail to notice the beautiful silver geraniums sparkling in the sunshine beside roads and trails. Called *nohuanu* or *hinahina* ("gray-gray") by the ancients, this low, rounded shrub is bedecked with white, five-petalled flowers (one inch across) and silvery, toothed leaves **(RIGHT)**. Specialized insects and spiders—also silvery—are so adapted to this geranium they never leave it. Silver geranium is also found in alpine life zones of the Big Island.

✚ SPATHIPHYLLUM

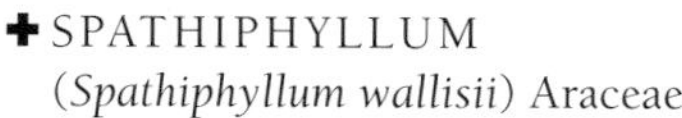

(*Spathiphyllum wallisii*) Araceae

Originally from Colombia, spathiphyllums evolved in the dark understories of lofty rainforests. Consequently, they can tolerate very low light levels. This is why we see spathiphyllums growing indoors where other ornamentals are only represented by their silk or plastic imitations. Spathiphyllums are less fussy about humidity and temperature controls than are their relatives, anthuriums. Under the watchful eye of hybridists, spathiphyllums now come in an array of flower and plant sizes.

✚ SPIDER LILY

(*Crinum asiaticum*) Liliaceae

Clear across the Pacific, spider lilies have always been popular ornamentals. They can survive, neglected, in hot sand, superannuated coconut plantations, and barren lava fields. Spider lily seeds are specially equipped with cork to aid in seawater dispersal. Even decades after homesites have been abandoned, their tough bulbs persist, sending aloft huge stalks of delicately spidery flowers year after year. Today, in Hawai'i's lowlands we still find spider lilies beautifying homes and public places in the old Pacific tradition. Like most lilies, their fragrance is superb, especially at night.

✚ STEPHANOTIS
{*Marsdenia (= Stephanotis) floribunda*}
Asclepediaceae

Another native of Madagascar (also called Madagascar jasmine or creeping tuberose), this ravishingly scented vine bears clusters of snow-white, tubular, waxy flowers. In Hawai'i, its association with weddings is so strong that the local name, *pua-male,* means "wedding flower". It is usually available at florists for decoration and corsages. Although a member of the milkweed family, its the flowers are quite different from typical milkweeds. Look for its large, mango-like fruits, which seem a little out of place on this non-bushy vine.

■ TETRAMALOPIUM, ALPINE
(*Tetramalopium humile haleakalae*) Asteraceae (Compositae)

This low, tufted "daisy", a specialty of Hawai'i's alpine and sub-alpine lava deserts, is familiar to high elevation hikers. The species occurs from 6000 to 10,000 feet in Haleakala Wilderness Area (Maui), and on Mauna Loa, Mauna Kea, Hualalai, and Kilauea (Big Island). Pictured is a subspecies with a highly restricted range, found only on East Maui. As in all daisies, the flowers are actually composite flowers, composed of ray florets (the outer "petals") and disk florets (the "center"), here white and pink, respectively. Tetramalopium often prefers microclimates close to chunks of lava or in the shade of larger boulders.

✚ TIARE OR TAHITIAN GARDENIA
(*Gardenia taitensis*) Rubiaceae

Native to many high island shores of the South Pacific, *tiare* has the distinction of being one of the few cultivated plants native to Polynesia. I love it so much we almost named one of our daughters Tiare. The intoxicatingly sweet-fragrant blossoms (two inches across) are unadorned works of art. A single flower is frequently used for personal adornment in islands beyond Hawai'i. Try adding one or two to warm coconut oil, as is the custom in Tahiti and Samoa, where *tiare* is cultivated extensively. Unfortunately it grows poorly upcountry, but thrives in moist lowlands.

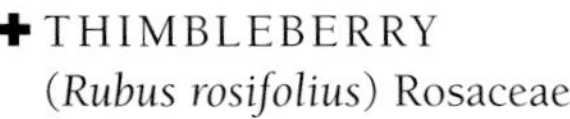

✚ THIMBLEBERRY
(*Rubus rosifolius*) Rosaceae

Native to Asia, this white-flowered, edible weed came to Hawai'i in the 1880s via Jamaica. Abundant in Hawai'i, it favors wet disturbed areas from about 100 to 5000 feet. Its thimble-sized, raspberry-like fruits are slightly insipid, but boil down into a good jam. Thimbleberry's stems are freely armed with small, recurved spines, so be prepared for scratched, bloody legs if you end up hiking through it in shorts.

✚ THUNBERGIA, LARGE-FLOWERED
(*Thunbergia grandiflora*) Acanthaceae

This striking vine trails flower-filled stems over balconies, trellises, and arbors. It may be blue or white. Native to India, large-flowered thunbergia now grows ornamentally in many tropical countries. Under ideal conditions, it climbs rampantly to 60 or more feet high. The leaves are variable in shape, but usually heart-shaped and angular.

✚ TREE HELIOTROPE
(*Tournefortia argentea*) Boraginaceae

(BELOW) A hardy coastal small tree, tree heliotrope is perfectly at home on island beaches, windward or leeward. Although it occurs naturally on practically every tropical Pacific island, its tiny seeds never reached Hawai'i on their own, so were introduced last century. Tree heliotrope's velvety leaves are crisply succulent, enabling the plant to withstand scorching heat and hypersaline sands.

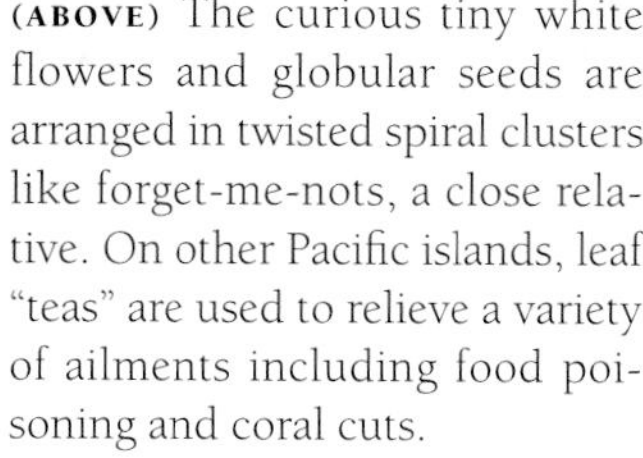

(ABOVE) The curious tiny white flowers and globular seeds are arranged in twisted spiral clusters like forget-me-nots, a close relative. On other Pacific islands, leaf "teas" are used to relieve a variety of ailments including food poisoning and coral cuts.

■ '*ULEI* OR HAWAIIAN HAWTHORN
(*Osteomeles anthyllidifolia*) Rosaceae

'Ulei, or Hawaiian hawthorn, is a hardy, sprawling evergreen, sometimes stiffly vine-like, at other times a bushy tree. It grows from coastal cliffs up into the alpine zone, tolerating incessant rain and prolonged droughts. *'Ulei's* white, plum blossom-like flowers are not especially "tropical", which likely relates to its temperate South American ancestry. In old Hawai'i, its purple berries provided a lavender infusion for dyeing *kapa* (bark cloth), and the flowers were woven into *haku* leis. The wood too, was used in unusual ways: fishnet hoops, bows for shooting rats, and a musical instrument called *'ukeke*.

Golden shower (*Cassia fistula*) is one of the parents of the equally beautiful hybrid rainbow shower (*Cassia fistula X C. javanica*).

YELLOW FLOWERS

Yellow flowers are abundant in Hawai'i all-year, from the seacoast to elevations up to 10,000 feet. An array of sparkling ornamentals includes orchids, gold-trees, shower trees, silver trumpet tree, wedelias, and heliconias, while natives include *koa*, greensword, *'ilima, hau,* and endemic hibiscus.

Yellow is a common color in flowers derived from both tropical and temperate regions. This is because birds, bees, and many other pollinating insects are attracted to its color. Even small amounts of yellow act as stimuli. No wonder stamens are yellow... it is the perfect color to excite all potential pollinators.

(RIGHT) A creative sunny arrangement incorporating day lilies, Oncidium orchids, and anthuriums.

✚ ALLAMANDA
(*Allamanda catharctica cv. 'Hendersonii'*) Apocynaceae

A widely cultivated ornamental vine from Brazil, allamanda's golden-yellow blossoms brighten Hawai'i's lowland gardens and resorts. The tubular, broad petalled flowers (three inches across) have deep throats and brown buds. Allamanda's Hawaiian name (rarely used), *lani-ali'i,* means "heavenly chief". This implies that after allamanda was introduced into Hawai'i it was recognized as a flower "fit for a king", since in olden times yellow and gold were exclusively royal colors.

▼ BANANA
✚ (*Musa* x *paradisiaca*) Musaceae

Although everyone is familiar with bananas, few take a moment to check out their attractive yellow flowers. Long and tubular, these bud out in successive curved layers, then ripen into "hands" of fruit. On every island, hikers in montane valleys may run across wild bananas, leftovers from former times when Hawaiians were more self-sufficient than today. These are generally the *'iholena* variety, a cooking banana with coppery tinged leaves. The fruits are starchy and puckery if eaten raw but excellent cooked. Pictured is the 'Williams' hybrid, commonly cultivated.

✚ BANKSIA, 'ORANGE FROST'
(*Banksia prionotes*) Proteaceae

Like most banksias, 'Orange Frost' hails from Western Australia, where its native habitat is semi-arid desert scrubland. Originally called "orange banksia", Maui's new commercial name focuses on its unusual "frosted" appearance. An easy grower, it grows both in farms and private gardens upcountry. Its compact, woolly buds open slowly to expose hundreds of orange-white spiky flowers in an ever-expanding band of fiery saffron. 'Orange Frosts' composite flowerheads and saw-toothed leaves last for months. The latter inspired the plant's botanical appelation, *prionotes,* literally "saw-toothed margins".

✚ BE-STILL TREE
(*Thevetia peruviana*) Apocynaceae

This shiny-leaved evergreen brightens roadsides, gardens and resorts throughout Hawai'i's lowlands. The odd common name derives from the fact that the leaves are always moving, even in the slightest breeze. Be-still's fragrant, funnel shaped flowers never seem to open fully, better suiting the Spanish name, *campanilla,* "small bell-flower". All parts of this plant, like oleander, containing a poisonous, digitalis-like drug called thevetin. Be-still was once grown commercially in Hawai'i for the manufacture and export of this drug.

✚ CANDLE BUSH
{*Senna (= Cassia) alata*} Leguminosae

A native of Central America, candle bush has long been grown in Hawaiian gardens, but is less common now than formerly. It is an unmistakable, sprawling bush (10-12 feet tall and wide), occasionally growing semi-wild along the roadside. The waxy bracts covering its flowerbuds inspired the Mexican name, *flor de secreto* ("secret flower"). Candle bush is an ancient Aztec medicinal plant introduced early to American pharmacology. Its leaves and seeds are beneficial in the treatment of ringworm, poisonous bites, and other skin ailments. Note the large, brown, winged seedpods (hence C. alata meaning "winged").

✚ DAY LILY

{*Hemerocallis lilio-asphodelus (= flava)*}
Liliaceae

Day lilies bloom most of the year at all elevations (see "Orange" chapter). Although long naturalized on the mainland, they have not yet escaped into the wild in Hawai'i. Day lily flowers and buds, rich in carotene (Vitamin A), impart a pleasing, *picante* (hot-spicy), taste to salads and stir-fried dishes. We once grew so many day lilies, I cooked up a cupful of the small root tubers as an experiment. They were crisp and delicious.

✚ EVENING PRIMROSE

(*Oenethera stricta*) Onagraceae

As you drive through alpine shrublands on Maui and the Big Island, you may notice an abundance of yellow "wildflowers." These are evening primroses, native to Chile and Argentina. Since they originated in a cool, temperate climate, they cannot survive in hot tropical lowlands. In Hawai'i they are consequently restricted to high elevations (4000 to 9000 feet). They take two years to complete their life cycle—the first year a basal leafy rosette develops and the second year flowers appear. True to their name, they open late in the day to expose pollen to twilight-flying insects.

Art Medeiros/Haleakala National Park

Cameron Kepler

✚ FENNEL

(*Foeniculum vulgare*) Apiaceae (Umbelliferae)

There are few places these days where you can collect herbs and spices by the roadside; Kula, Maui is one. Fennel is a dill-like herb, a few feet tall, with anise-scented, delicately dissected leaves and pale yellow flowers reminiscent of Queen Anne's lace. It grows abundantly from about 3000 to 4000 feet in upcountry Maui. Native to Europe, fennel's leaves, stem, flowers, and seeds have long been utilized to improve vision, "make people lean that are too fat", and correct flatulence. Don't forget fennel in your Indian curries and seafood dishes.

✚ GAZANIA OR TREASURE FLOWERS
(*Gazania spp.*) Asteraceae (Compositae)

Gazanias (see also "White" chapter) are favorite ground covers in upland residential areas. They are easy to spot: yellow daisy-like flowers with a beady ring of brown dots around the central disc. The bright yellow color attracts potential pollinating insects to the flower, while the dark dots specifically guide them to nectar and pollen supplies. Gazania leaves are elliptical, with a dense cover of white hairs, imparting a silvery appearance. Their native habitat in South Africa must be dry, alpine, or both. On rainy or dark days gazania flowers are reticent to open; they also close at night. Don't bother using them for leis.

✚ GINGERS
(*Hedychium, Zingiber*) Zingiberaceae

(RIGHT) A relative newcomer to Hawai'i, golden beehive ginger (*Z. spectabile*) is a spectacular ginger, whose scientific name means exactly that. It hails from Asia, where millions of years of evolution has produced an astonishing 1200 species of gingers. Layers of ridged scallops, intricately molded from waxy tissues, comprise the golden beehive's eight-inch flowerhead. Small yellow and purplish flowers, speckled and orchid-like, peek out from inside the scalloped bracts. Golden beehives are grown commercially in windward Maui and the Big Island. Watch for them in flower arrangements in resorts and public places.

(LEFT) *'Awapuhi-melemele* (literally "yellow ginger"), along with its cousin, white ginger, grows luxuriantly along in the lower montane forests on all islands except Lanai. At times, its impenetrable entanglements challenge even the hardiest hikers. However, its flowers are beautiful even if naturalized plants are a pest. Ginger flowers are unusual because the petals come in different shapes (slender, or broad and notched like a pair of closed wings). To further confuse the novice, the two egg-shaped "petals" are actually modified male reproductive parts (staminodes).

Kahili ginger (*H. gardnerianum*), another Himalayan beauty, was first collected in Hawai'i around 1940. The cylindrical form and enchanting fragrance of its regal flowerhead, are hard to forget. Enjoy *kahili* ginger in the islands, since the flowers are too fragile to be shipped in boxes. The name honors Hawaiian *kahili* or royal standards. These familiar insignia of bygone days—red and yellow cylinders made from bird feathers atop long shiny poles—were indispensible on ceremonial occasions. Beautiful in gardens, *kahili* ginger becomes a forest nuisance when it escapes, for example, Glenwood—Volcano (Big Island), Kokee (Kauai), Hana-Kipahulu (Maui), since it crowds out all other seedlings.

■ GREENSWORDS

(*Argyroxiphium grayanum*) Asteraceae (Compositae)

This is one of Hawai'i's exceptional endemics: it is not only restricted to Maui, occuring only within selected boggy cloud forest habitats, it is rare, vulnerable, and endangered. Its closest relative (*A. virescens*) has not been seen since 1945. **(LEFT)** Typical East Maui greensword habitat, an extremely fragile bog at 6100 feet elevation. **(RIGHT)** A greensword flowerhead, showing dozens of yellow daisies. Although its general form resembles that of its cousin, the silversword, the greensword never develops hairy, succulent leaves, since its soggy habitat precludes any need to conserve water.

(LEFT) Because of the greensword's largely inaccessible habitats, it is difficult to see them in the wild. However, the National Park Service sometimes grow them for display: this beauty in 1985 was 12 feet high. **(BELOW)** Young greenswords from the Violet Lake bog, West Maui. The volcanic side-peak is Puu Nakalalua (4503 feet), high above Kaanapali and Kapalua. West Maui greenswords are smaller than their Haleakala counterparts.

▼ *HAU,* A PACIFIC-WIDE COASTAL TREE
◆? (*Hibiscus tiliaceus*) Malvaceae

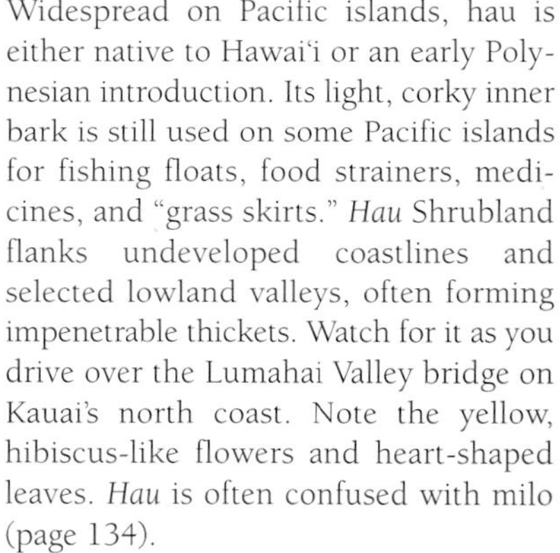

Widespread on Pacific islands, hau is either native to Hawai'i or an early Polynesian introduction. Its light, corky inner bark is still used on some Pacific islands for fishing floats, food strainers, medicines, and "grass skirts." *Hau* Shrubland flanks undeveloped coastlines and selected lowland valleys, often forming impenetrable thickets. Watch for it as you drive over the Lumahai Valley bridge on Kauai's north coast. Note the yellow, hibiscus-like flowers and heart-shaped leaves. *Hau* is often confused with milo (page 134).

✚ HELICONIAS
(*Heliconia*) Heliconiaceae

(RIGHT) Everblooming parrot's beak heliconia (*H. psittacorum*), a popular ground cover, comes in a stunning array of varieties, all less than four feet high. The "painted" flowerheads, glow as if lit internally (see also "Orange" chapter). Parrot's beaks are versatile for homeowners, flourishing in hot humid lowlands, in drier, well-watered sites, and even at medium elevations. Commercial plots of 10 square feet average more than 200 blooms annually. For longer lasting blooms, cut *before 8 a.m.*, and every few days change the water, add a pinch of sugar, and slice off the stem ends.

(LEFT) Fuzzy yellow hanging heliconia (*H. xanthovillosa*) is a newcomer from tropical America, presently only in Hana (Maui), but with much promise for the future. The entire flower stalk and boat-like bracts are covered with soft, long, golden hairs, inviting one to stroke it like a cat! **(BELOW)** Dwarf golden heliconia (*H. aurantiaca*) bears five inch-long orange flowerheads packed with yellow, banana-like florets. Its leaves have a lacquered sheen.

■ HIBISCUS, NATIVE AND INTRODUCED
✚ (*Hibiscus*) Malvaceae

Maui's pure yellow hibiscus are of particular interest. **(CENTER LEFT)** X Hibiscus calyphyllus was, until the late 1980s, called Rock's Kauai hibiscus (H. rockii), designated as endemic to Kauai. A lovely yellow hibiscus, it joined the ranks of Hawai'i's beautiful natives. What a blow when botanists discovered that it is really an African hibiscus which escaped from cultivation decades ago! Despite its botanical history, the landscapers have now discovered it! **(BOTTOM RIGHT)** An introduced hybrid hibiscus (X *H. rosa-sinensis* 'Thunderball'), characterised by large overlapping, crepy yellow petals, with a red eye.

(BOTTOM LEFT) A rare, localized native, ma'o hau hele (*H. brackenridgei*), pronounced "*ma-oh-how-hell-ley*", is a sprawling shrub, officially endangered. Pictured is the extremely rare Lanai form, which now dots private and public gardens. In 1988 this bright yellow hibiscus became Hawai'i's official state flower, replacing *kokio-'ula* ("Red" chapter).

✚ IXORA
(*Ixora chinensis*) Rubiaceae

Occasionally you will run across an ixora bush with rich cream, star-like flowers instead of scarlet (p. 87). Variants and cultivars come in many sizes and colors these days: pink, salmon, white, cream, and orange, as well as the parental red. If the flower clusters are particularly large and if they emit a sweet, heavenly scent, then you have the uncommon sweet ixora (*Ixora odorata*) from Madagascar. Its wood is extremely hard.

✚ JASMINE, ITALIAN YELLOW
(*Jasminum humile*) Oleaceae

Easily recognized, this dense, rambling vine bears stiff, angled stems and small numbers of yellow, scentless flowers (one-half inch across). Although called "Italian jasmine", it is native to the Himalayan region. As expected, it is fairly cold tolerant; look for it in mid- to higher elevation residential areas such as Kokee (Kauai), Kula (Maui), and Volcano (Big Island).

◆ *'ILIMA*, A NOTABLE LEI FLOWER
(*Sida fallax*) Malvaceae

Distributed across the Pacific and in China, 'ilima, a miniature hibiscus, once held the title of Hawai'i's state flower before its larger, reddish cousin (*kokio 'ula*) took over in 1923. *'Ilima* is now the floral symbol of Oahu. Watch for it in wastelands and coastlines such as Kaena Point (Oahu), Wailea Point (Maui), southern Kauai and Molokai. It is honored by centuries of love from generations of Hawaiians; *'ilima* leis have always been special, and the plant had medicinal and other uses too. Since each lei requires around 500 blossoms, today's trend is to replace fresh flower by circles of closely-strung crepe paper, which children learn early in their schooling.

(LEFT) The golden flowered hairy abutilon or *ma'o* (*Abutilon grandifolium,* pron. "mah-oh"), a weed, is also used as an 'ilima substitute. **(RIGHT)** If you are ever presented with an authentic *'ilima* lei, bask in this almost unparalleled tribute of *aloha.*

✚*KIAWE* OR MESQUITE

(*Prosopis pallida*) Leguminosae (Fabaceae)

(BELOW) Because the tiny-leaved, spiniferous *kiawe* (pron. "kee-ah-vee" dominates Hawai'i's leeward lowlands, many people assume that it is native. Unfortunately it has *replaced* precious *koa, 'ohi'a, wiliwili, naio,* and numerous other native trees, shrubs and herbs, many of which are now rare or endangered. Native to the Andes, South America, *kiawe* was introduced in 1828 by a Roman Catholic missionary from the Royal Gardens in Paris.

(ABOVE) Once *kiawe's* protein-rich seedpods were collected in abundance for cattle feed. Its flower spikes are less known than its accursed spines, frequently encountered on natural beaches, pastures, and in coastal areas on all islands.

■*KOA*, A MAJESTIC HARDWOOD

(*Acacia koa*) Mimosaceae (Leguminosae)

Koa, a special Hawaiian tree. Biological research indicates that it is a true *keiki* o ka 'aina ("child of Hawai'i's earth") of millions of years' standing, perhaps the most ancient tree in the islands. Dozens of endemic insects and several Hawaiian honeycreepers are (and were) associated primarily with *koa,* for example, the rare Maui Parrotbill (*Pseudonestor xanthophrys*) and the Big Island's extinct Greater *Koa*-Finch (*Rhodacanthis palmeri*). Mature *koa* foliage resembles that of eucalypts, although they are unrelated; both independently evolved similar strategies to survive in dry environments. *Koa's* sickle-shaped leaves are not true leaves but expanded, flattened leafstalks called phyllodes. Young leaves are fern-like.

Cameron Kepler

(LEFT) Watch for *koa* trees as you hike anywhere in dry or semi-moist (mesic) forests, especially on the Big Island and Kauai (Waimea Canyon, Kokee). Watch for saplings to observe the young, fern-like leaves. Koa Dry Forest formerly covered vast acreages on the dry to moist slopes on all islands, but was mostly cleared for cattle. In spring, the trees burst forth with masses of creamy yellow pompons.

✚*KOA*, FORMOSAN
(*Acacia confusa*) Leguminosae

As its name suggests, Formosan *koa* is native to Taiwan (formerly Formosa) and also to the Philippines. Introduced to Hawai'i in 1915, Formosan koa is a timber tree that has become naturalized in moist to wet lowland forests. A good area is on Maui along the Hana Highway (Kaumahina State Park, Haiku). A medium to tall tree, it is easily distinguished from the authentic Hawaiian koa by golden, rather than cream or yellow flowers, and narrower, shorter, and straighter "leaves" (actually flattened leafstalks).

■ *KUPA'OA*, A WILDERNESS AREA SPECIALTY
(*Raillardia menziesii*) Asteraceae (Compositae)

One of the scant bushes adapted for survival in the dry cinders of Haleakala is *kupa'oa*. A Maui endemic, it is especially common above 9000 feet. Peak blooming is between July and December. Note the daisy-like flowers and leathery leaves ranked in fours. This tight, regular leaf arrangement is characteristic of many alpine plants, since it aids water retention. Although *kupa'oa* looks extremely different from the silversword, remarkable hybrids between the two species have been observed and studied.

Ron Nagata/Haleakala National Park

✚LOLLIPOP PLANT

(*Pachystachys lutea*) Acanthaceae

Lollipop plant, grown in Hawai'i since the 1970s, is a distinctive, short shrub that blooms abundantly, providing neat splashes of color (sun or part-shade) in borders, hedges, or in containers. It is very similar to its close relative, the yellow shrimp plant (see common red variety of shrimp plant, "Red" chapter). However, the distinguishing characters are easy: the lollipop's flowerheads are always short and erect, rather than arching, and the leaves are twice as big (to 6 or 7 inches, rather than half that in the shrimp plant).

■*MAMANE*, "THE PALILA TREE"

(*Sophora chrysophylla*) Papilionaceae (Leguminosae)

Golden, pea-like blossoms, long knobby seedpods, and lacy evergreen foliage characterize Hawai'i's native *mamane*. Feral goats and sheep have decimated large areas of mamane on the Big Island; old-timers may remember the heated court case relating to the endangered Hawaiian honeycreeper, the Palila. *Mamane* Forest is also relictual on East Maui: the best place is Haleakala National Park, where rangers continue active goat management. Watch for sizable patches of *mamane* beside the road en route to the volcano rim. Optimal blooming time is spring.

■*MA'O* OR HAWAIIAN COTTON

(*Gossypium tomentosum*) Malvaceae

Ma'o (pron. "mah-oh"), extremely uncommon, is native to Maui. It was formerly the dominant species of Ma'o Shrubland, a native habitat type occurring on arid, rocky coastal sites. One of the relatively few Hawaiian endemics derived from tropical America, it is rare in the wild but flourishes at Wailea Point. Notice the small amount of brown fluff and soft, downy, maple-shaped leaves. Recently, breeders have realized *ma'o's* genetic potential to improve agricultural cotton: it produces a hardy strain that is highly resistant to aphids and boll weevils. This is due the flowers' scant nectaries—they offer little or no sugary reward to visiting insects, who consequently stay away!

▼ *MILO*, A TRANS-PACIFIC COASTAL HARDWOOD
◆ (*Thespesia populnea*) Malvaceae

The richly colored wood of *milo* (portia tree) is still esteemed in Hawai'i. To my mind, its swirled patterned layers of cream, fawn, tan, chestnut, and deep pink surpass monkeypod. Although widespread on Pacific islands, *milo* probably did not reach Hawai'i without the help of voyaging Polynesians. It is commonly seen wild or planted along coasts. Many people confuse it with *hau,* but they are easily distinguished. Note the darker, shiny, very pointed leaves of milo, seedpods that do not split open, and a tree-like habit. In contrast, *hau* has dull, wide heart-shaped leaves, seedpods that split, and a rambling, shrubby habit.

■ *NEHE* (*Lipochaeta*) and *KO'OKO'OLAU* (*Bidens*), NATIVE DAISIES
Asteraceae (Compositae)

These two groups of native yellow daisies are combined here because they are similar, often occur together, and form groups of about 20 endemic Hawaiian species each. Many are rare. **(LEFT)** The endangered *Lipochaeta lavarum* is found in dry areas of all islands except Oahu and Kauai. Its cheery flowers and whitish leaves peep up from bone dry, clinkery lava. **(RIGHT)** *L. integrifolia,* another endangered, small-leaved *nehe,* is today found only on scattered windward coasts of all the main islands.

(ABOVE) *Bidens hillebrandiana* frequents the windward coasts of Molokai, Maui, and northeast Big Island. Its leaves are waxy and semi-succulent. **(RIGHT)** The rare *B. mauiensis* dots coastal bluffs and arid shrublands of Maui, Lanai, and Kahoolawe.

✚ ORCHID, EPIDENDRUM
(*Epidendrum x obrienianum*) Orchidaceae

Scarlet epidendrums, of varying colors and encountered upcountry, have been discussed elsewhere ("Orange", "Purple", "Red" chapters). They are easy to grow, forming large clusters after several years. Epidendrums characteristically produce mases of fluffy seeds in their fat seedpods, as well as *keikis* (tiny plantlets) on the abre stems. The development of plantlets is rather unusual in this family, since orchid flowers have evolved for insect pollination—in some cases both orchids and insects are elaborate and very species-specific.

◆ PEA, BEACH
(*Vigna marina*) Papilionaceae (Leguminosae)

This gay little native pea (*nanea*), twining around *hala* roots **(LEFT)**, crawling over seashore boulders, and binding sand in many coves, reminds us that Hawai'i is a truly oceanic archipelago. Over eons of time, with *pohuehue* (see "Purple" chapter) beach pea has helped shape Pacific atolls and islands by binding sand and helping to create soil. Few ancient Hawaiian uses are known, but *nanea's* medicinal uses on other Pacific islands include eyedrops and infusions to bathe fractures and other skin ailments. In Samoa and Tonga it is particularly effective as a "spirit" or "ghost" medicine.

✚ PARTRIDGE PEA
{(*Chamaecrista nictitans* (= *Cassia lechenaultia*)}
Fabaceae (Leguminosae)

A rather attractive weed, partridge pea is widely distributed along roadsides and in dry or moist wastelands. It often occurs with sensitive plant (see "Pink" flowers). Originally from tropical America, it has been introduced into many tropical areas. Notice the feathery leaves and yellow, pea-like flowers which usually open in ones or twos. Partridge peas, like most "sweet pea-type" flowers, are pollinated by bees, bumblebees, and butterflies.

✚ PLUMERIAS
(*Plumeria rubra*) Apocynaceae

The first plumeria plant ("common yellow") in Hawai'i was brought in 1860 by a famous early botanist, William Hillebrand. Native to Mexico and Guatemala, it bears enticingly fragrant flowers with yellow centers. Incidentally, there are two yellow-and-white-flowered species (see "White" chapter): *Plumeria rubra* has *deciduous, oval, pointed* leaves and numerous flower colors; and the Singapore plumeria (P. obtusa), has *larger, more oblong, evergreen* leaves with a *blunt tip*. In recent years, controlled plumeria hybridization has been carried out at the University of Hawai'i, but most of the tremendous variation in size, color, and perfume in Hawaiian plumeria trees is derived from natural hybridization.

■ PORTULACA, MOLOKINI
(*Portulaca molokiniensis*) Portulacaceae

Portulacas (purslanes), with their rounded succulent leaves and distinctive, rose- or buttercup-like flowers, are well-known. On Maui, rose moss *(P. grandiflora)*, with a variety of flower colors, enlivens sunny gardens (see "Purple" chapter), and pigweed *(P. oleracea)*, with small yellow flowers, grows along roadsides. **(LEFT)** A new species of 'ihi *(P. molokiniensis)*, restricted to the volcanic slopes of Molokini Island and two coastal locations on Kahoolawe, was discovered in the early 1980s by Maui forester/botanist, Bob Hobdy. **(BELOW)** The entire geographic range of the rare Molokini portulaca: tiny, crescentic Molokini and larger Kahoolawe, its natural habitats, and Wailea Point, where it is cultivated as a conservation measure and for the public's enjoyment.

✚ PROTEAS
(*Protea, Leucospermum*) Proteaceae

(BELOW) The University of Hawai'i's Agricultural Experiment Station in Kula, Maui, is the world's leading center for protea research. Their first released hybrid in the 1980s, was 'Hawai'i Gold' (*Leucospermum* 'Hawai'i Gold'), a natural genetic mix of two pincushion proteas (*L. cuneiforme, L. conocarpodendron*). When this unexpected, pure gold flower appeared on Maui, Dr. Philip Parvin traced its origins and found that a South African sunbird (an ecological equivalent of a hummingbird) had brushed pollen from one species onto the stigma of another, producing an abnormally viable hybrid. Now propagated by cuttings, 'Hawai'i Gold's' flowers, adorned with furry, gold loopings and matching "pins" begin blooming in February and continue for several months.

(ABOVE) Clad in layered, woolly clothing, inside and out, the glorious queen protea (*P. magnifica*) is one of Hawai'i's commercially grown specialties. Appropriately named "magnificent protea", it is normally pink but also comes in red, orange, and chartreuse. Its spangled, furry adornments may be snowy-white or gold, but the velvety central knob is always black.

✚ ROSE APPLE
(*Syzygium jambos*) Myrtaceae

Native to southeast Asia, rose apple came to Hawai'i via Brazil in 1825 on the frigate *Blonde*. Although originally conceived as garden ornamentation, it is now primarily found naturalized in wet, windward forests **(RIGHT)**. Although rose apple's yellow, pompon flowers resemble those of yellow *'ohi'a* (red 'ohi'a, p. 89), note that its leaves are long and slender rather than short and round. The unusual fruits are a little dry and fibrous, but *taste the way roses smell.*

✚ SHOWER TREE, GOLDEN
(*Cassia fistula*) Fabaceae (Leguminosae)

(**RIGHT & CHAPTER HEADING**) Grape-like clusters of golden "sweet peas" sparkle in August sunshine. Look for them in established towns (Honolulu, Kokee, Lahaina, Kaunakakai). Their long seedpods (one to two feet) bear ample seeds for leis and new seedlings, but some consider them messy. The modern trend is to replace golden showers with the equally beautiful rainbow shower (**BELOW** and in "Orange" chapter), which blooms even more prolifically and produces few or no seedpods. An old name for golden shower is "purging cassia", since it was one of the several species of Cassia whose pulp pods were used for intestinal cleansing. (Today, commercial senna is made from a related species, appropriately C. senna.)

✚ SHOWER TREE, RAINBOW
(*Cassia* x *nealae*) Fabaceae (Leguminosae)

The official tree of Honolulu—where there are thousands—the rainbow shower is popular in the newer commercial and residential areas on all islands. All shower trees produce fine quality timber and medicinal ingredients. Rainbow shower trees are truly splendiferous. Every tree is different, a character which stems from their hybrid nature: a cross between golden and pink-and-white showers (see above and "Pink" chapter). At close range, the rainbow shower's overall yellow appearance separates into cream, bronze, apricot, pink, rose, and gold. Some, for example, at the University of Hawai'i in Honolulu, are a delicate yellow. (See also p. 39.)

✚ SILK OAK
(*Grevillea robusta*) Proteaceae

Originally from Australia, the silk oak came to Hawai'i around 1880 as a timber tree. More than two million seedlings were planted. Under optimum conditions, trees grow 100 feet tall. Japanese White-eyes, small green birds with bright white "spectacles" and tubular tongues, often flock around silk oaks on sunny mornings, sucking nectar and foraging on insects. (**RIGHT**) An eye-catching, unusual lei utilizes silk oak, anthurium, *kamani* seeds, fan palm, baby palm seeds, and ferns.

Silk oak hails from a dry temperate climate, therefore thrives best at mid- to high elevations, for example, Kamuela (Big Island) and Kula (Maui). **(LEFT)** Its feathery, silky-silvery foliage glistens in the sunshine. **(BELOW)** Peak blooming is in late spring and summer, when spidery, tooth-brush-like flowerheads (3-5 inches long) appear in profusion.

✚ SPRAY-OF-GOLD

(*Galphimia gracilis*) Malpighiaceae

Gay clusters of small, star-like flowers characterize this medium-sized shrub, found in gardens and around resorts. The Mexican name, *lluvia de oro* ("rain of gold") is also descriptive. The family to which it belongs is a little-known tropical family which includes acerola fruit (extremely high in vitamin C), Singapore holly, and various medicinal plants. Most flowers bear petals which are pinched in at their bases.

✚ TOBACCO, WILD

(*Nicotiana glauca*) Solanaceae

Wild, or tree, tobacco, is a scraggly, "open" bush with long slender branches. Native to Uruguay and Argentina, it thrives in very dry areas, for example, Kihei and Wailea (Maui). The flowers are narrow yellow tubes (1 1/2 inches), ending in five small rounded petal lobes, while the smooth, white-waxy (glaucous) leaves are arranged alternately along the stems. Wild tobacco is related to tobacco (*Nicotiana tabacum*), both of which were named in honor of J. Nicot, a French Ambassador to Spain, who sent tobacco seeds to France in 1560.

✚ TRUMPET TREE, SILVER

{*Tabebuia aurea (= T.argentea)*} Bignoniaceae

(RIGHT) Silver trumpet tree is a relative newcomer to Hawai'i, currently common in commercial, street, resort, and garden landscaping. It is easy to recognize but difficult to find in books. Note its *leathery, elliptical, whitish leaves, arranged like the fingers of a hand (palmate),* and clusters of golden, tubular flowers with five spreading lobes. It can be pruned into radical shapes as a specimen tree.

(LEFT) Some people call the silver trumpet tree "yellow tecoma", but this is a misnomer. Yellow tecoma is another tree altogether, uncommon in Hawai'i. It bears similar yellow flowers but has *opposite, very pointed, green, pinnate leaves (divided like a feather) which are margined with distinct teeth and are not leathery.*

✚ WATTLE, BLACK

{*Acacia mearnsii (= A.decurrens), A. paramattensis*} Fabaceae (Leguminosae)

(RIGHT) Black wattles are much less attractive than hundreds of Australian wattles ... but they yield valuable tannins for tanning leather, and were introduced for that purpose. The rusty-brown, woody bark contains a high proportion of tannin (to 42%) and is very astringent. In Hawai'i, one tree can yield up to a quarter-ton of bark. The 2 species of black wattles are abundant in Kula, Maui **(RIGHT)** where the dark foliage, finely feathery leaves, and fragrant, dirty-lemon pompons (similar to *koa* flowers) are known to all. Fast growing, the trees are used for firewood on all the main islands up to 5000 feet elevation.

✚ WEDELIA

(*Wedelia trilobata*) Asteraceae (Compositae)

(LEFT) A ubiquitous ground cover, wedelia cannot be missed around resorts, public buildings, condominiums, private gardens, and naturalized along highways and in state parks. Native to the American tropics, it grows rapidly by creeping and rooting. Fortunately it sets few seeds in Hawai'i, thus its reproductive capacity is essentially limited to vegetative growth; hopefully it will not become a serious pest. Wedelia superficially resembles some of the native yellow daisies such as *nehe* (*Lipochaeta*) and *ko'oko'olau* (*Bidens*), which are much rarer (p. 134).

INDEX OF ENGLISH NAMES

INDEX OF HAWAIIAN NAMES

INDEX OF SCIENTIFIC NAMES

ABOUT THE AUTHOR

DR. ANGELA KAY KEPLER, a naturalized New Zealander, was born in Australia in 1943. A terrestrial ecologist and photographer, she writes books on Hawaiian and Pacific natural history, and surveys seabirds and terrestrial ecosystems on little-visited Pacific coral islands (Pan-Pacific Ecological Consulting, Inc.). She also works as a naturalist-lecturer on small ecotourism cruise ships. Kay holds degrees from the University of Canterbury (New Zealand), University of Hawai'i (Honolulu), and Cornell University (New York). She also spent one year as a post-doctoral student at Oxford University, England, and lived for three years in a rain forest in Puerto Rico.

Kay first came to Hawai'i as an East-West Center foreign student in 1964. Over the last 27 years she has authored (or co-authored) 12 books and numerous scientific publications, written newspaper columns on biological and cultural aspects of the Hawaiian Islands, and contributed regular articles and photos to island publications. She and her husband, Cameron, separately and together, have conducted forest bird and plant surveys, seabird studies, and endangered species research in Hawai'i and other Pacific islands, the mainland U.S., Alaska, the West Indies, and New Zealand. Their contributions to conservation have been numerous: in Hawai'i these include pivotal roles in Maui's Waikamoi Preserve and Molokai's Olokui Preserve.

The Keplers lived on Maui from 1977 to 1988, during which time they hiked practically every valley and mountain slope, surveyed almost every offshore island, and canoed many portions of shoreline. On Maui Kay discovered two new species of lobelias and unearthed a previously unknown extinct bird species (a rail). The Keplers also co-discovered a new bird species (Elfin Woods Warbler, *Dendroica angelae*) in Puerto Rico.

In 1988 the Keplers were transferred to the University of Georgia, Athens, GA. Kay continues work in the Pacific, and is still one of Hawai'i's top-selling authors.